Methods of Estimating Reserves of Crude Oil, Natural Gas, and Natural Gas Liquids

Methods of Estimating Reserves of Crude Oil, Natural Gas, and Natural Gas Liquids, first published in 1965, aims to throw new light on a field of knowledge vital to consideration of problems of public policy regarding future sources of energy. This book will be of interest to students of environmental studies.

Methods of Estimating Reserves of Crude Oil, Natural Gas, and Natural Gas Liquids

Wallace F. Lovejoy and Paul T. Homan

RFF PRESS
RESOURCES FOR THE FUTURE

First published in 1965
by Resources for the Future, Inc.

This edition first published in 2015 by Routledge
2 Park Square, Milton Park, Abingdon, Oxon, OX14 4RN
and by Routledge
711 Third Avenue, New York, NY 10017

Routledge is an imprint of the Taylor & Francis Group, an informa business

Publisher's Note
The publisher has gone to great lengths to ensure the quality of this reprint but points out that some imperfections in the original copies may be apparent.

Disclaimer
The publisher has made every effort to trace copyright holders and welcomes correspondence from those they have been unable to contact.

A Library of Congress record exists under LC control number: 65024790

ISBN 13: 978-1-138-85630-1 (hbk)
ISBN 13: 978-1-315-71912-2 (ebk)

Methods of Estimating Reserves of Crude Oil, Natural Gas, and Natural Gas Liquids

by

WALLACE F. LOVEJOY

and

PAUL T. HOMAN

RESOURCES FOR THE FUTURE, INC.

Distributed by

THE JOHNS HOPKINS PRESS,
BALTIMORE, MARYLAND 21218

Resources for the Future is a non-profit corporation for research and education in the
development, conservation, and use of natural resources. It was established in 1952 with
the co-operation of the Ford Foundation and its activities since then have been financed
by grants from the Foundation. Part of the work of Resources for the Future is carried
out by its resident staff, part supported by grants to universities and other non-profit
organizations. Unless otherwise stated, interpretations and conclusions in RFF publications are those of the authors; the organization takes responsibility for the selection of
significant subjects for study, the competence of the researchers, and their freedom of
inquiry.

This book is one of RFF's studies in energy and mineral resources, which are directed
by Sam H. Schurr. The research was supported by a grant to Southern Methodist
University. Wallace F. Lovejoy is professor of economics, and Paul T. Homan research
associate and former director of graduate studies, Department of Economics, at the
University. The manuscript was edited by Doris L. Morton.

Director of RFF publications, Henry Jarrett; *editor,* Vera W. Dodds; *associate editor,*
Nora E. Roots.

Foreword

The concepts and approaches employed in estimating the future availability of petroleum have led to considerable misunderstanding. They have also left substantial gaps in the quantitative information. In sponsoring the present study, Resources for the Future hoped that it would throw new light on a field of knowledge vital to consideration of problems of public policy regarding future sources of energy.

Professor Lovejoy and Professor Homan have written a deceptively simple book on the subject. Only those who have themselves been forced to labor through the raw information will recognize how much effort and hard thought have gone into this lucid account of the different reserves estimates and the approaches underlying them. Beginning with the concepts and procedures used by the American Petroleum Institute in producing its estimates of proved oil reserves (which are, in a sense, the "official" figures on the subject), the authors lead us through a number of other significant approaches which have been utilized, or suggested, to measure the nation's position in natural resources of crude oil, natural gas, and natural gas liquids.

The past instances in which the available statistics on oil and gas reserves have apparently yielded "wrong" answers on questions of the future adequacy of resources are so numerous that much skepticism now exists concerning the real worth of existing information. To a degree, as the authors point out, the improper application of reserves figures to deal with questions for which they were never intended has been the source of much of the difficulty.

Nevertheless, even when the underlying concepts and methods are fully understood, serious deficiencies still exist, because the statistics now being compiled are inadequate for dealing with important questions affecting the public interest. Consequently, at various points in the study, but mainly in the last chapter, the authors discuss the types of improvements in the figures which are needed and which appear to be within reach as a logical extension of the present programs of data collection and analysis.

It is our hope that the publication of this study will help to minimize the future misuse of reserves statistics, and that it will also call attention to the possibilities that exist for making improvements in the statistics. The attention of the petroleum industry, government agencies and the interested public is directed to both of these objectives.

SAM H. SCHURR
Director, Energy and
Mineral Resources Program,
Resources for the Future, Inc.

Preface

The present study is the lineal offspring of an earlier effort by the same authors. To establish the origin and purpose of the present study, we quote from the Preface of the earlier one.

In the spring of 1961, Resources for the Future, Inc. made a grant to the Department of Economics and the School of Law at Southern Methodist University to hold a seminar on economic and legal aspects of the petroleum industry. A central purpose of the project was to bring face to face around the conference table, for discussion of some fundamental topic, people from within the industry, academic personnel engaged in research upon the industry, and other persons who in a consulting or regulatory capacity were concerned with the problems of the industry. The initial project was conceived of as possibly the opening step in a series of studies in petroleum economics and law; and it was felt that the establishment of direct lines of communication between the various types of personnel concerned with the problems of the industry would be valuable.

It was decided, for purposes of this experiment, to choose a topic of limited scope and technical character. This limitation excluded consideration of, and argument about, controversial questions of policy. Following this principle, the subject chosen was "Oil and Gas Finding, Development, and Producing Costs." Within this limited scope, the orientation was not toward the specific costing problems of individual companies, but rather toward questions of whether meaningful cost studies could be made, whether existing cost concepts and methods of analysis are correct and useful, and whether cost data are essential information in evaluating the future availability of petroleum supplies.

The seminar met at Southern Methodist University in five sessions over a period of two and a half days, March 22 to 24, 1962. In preparation for the seminar the present authors prepared a background paper which was distributed in advance and served as the point of departure for the seminar discussions. This paper was revised and published, together with a summary account of the discussions.[1] Since circumstances did not permit original empirical research into actual cost figures, the study primarily centered around (1) the conceptual basis of cost analysis, (2) a review of earlier cost and availability studies, and (3) the bearing of costs upon regulatory activities and policy making.

In the course of the seminar, an extended discussion of the concept of "replacement cost" took place. This necessarily led into the methods of estimating reserves and the meaning of the estimates, since the replacement cost of oil, as usually calculated, relates finding and development costs to the proved reserve added annually, as estimated by the Committee on Petroleum Reserves of the American Petroleum Institute. Readers who wish to follow the whole discussion will have to consult the original, but the following passage from the *Summary*[2] is especially pertinent:

As the preceding part of this report has shown, the discussion of replacement costs necessarily included some discussion of petroleum reserve data—how these data are collected, aggregated, reported, and transmuted into a cost context. The significance of reserves is evident when it is recalled that finding costs per barrel and development costs per barrel are computed by dividing annual expenditures on finding and development by the barrels of additional reserves "proved up" during that year. Even if economic cost concepts (or some variation of these) are used, it is necessary to have a denominator for the fraction to obtain a cost *per barrel* of reserves. The discussion with respect to reserves was interspersed in the general discussion on costs; it is dealt with separately at this point in order to emphasize some of the unique aspects of reserve estimation. Three principal questions were as follows. How are reserve data collected? Are the data reliable and adequate for the uses to which they are put? Can reserve data be broken down by functional stages—e.g., discoveries, extensions, revisions—in a meaningful way?

The discussion brought out the procedures used by the American Petroleum Institute, the American Gas Association and the National Petroleum Council to collect and report reserve data. Most participants indicated that the annual breakdown between additions to reserves from discoveries on the one hand and from extensions and

[1] Wallace F. Lovejoy and Paul T. Homan, with Charles O. Galvin, *Cost Analysis in the Petroleum Industry* (Dallas: Southern Methodist University Press, 1964).
[2] *Ibid.*, pp. 110-12.

revisions on the other hand was not reliable enough for purposes of
cost analysis. Dividing total development costs by extensions and
revisions to get a development cost per barrel was not justified, as
there is no established relationship between the two. The same general
comment applies to finding costs and new reserves from discoveries.
It applies also if combined finding and development costs are related
to total additions to reserves.

The question was raised as to what extent there is a bias in the
official A.P.I. and A.G.A. reserve data. It was recognized that they
are, by their very definition, conservative, representing a working
inventory concept and not a prediction of probable or possible
recovery from known reservoirs. Companies customarily make
further estimates of probable reserves, although these figures contain
a large judgment factor and are apt to diverge widely even within a
company among different people. Some in the group advocated that
companies pool the estimates on the probable reserves or that the
A.P.I. and A.G.A. report both "proved" and "probable" reserves
by whatever methods are available to them. The group seemed to
agree that A.P.I.-A.G.A. estimates of reserves added by discoveries
for single years do not present a time series of much use for charting
the time trend of discovery.

Some members of the group emphasized a need for greater detail
in the reporting of reserves, especially in differentiating revisions and
extensions. This led to the further suggestion that it would be valuable
if specific additions to reserves could be attributed to specific factors,
such as changes in information, further drilling or recompleting,
introduction of secondary recovery operations, technological innovations or applications, and changes in underlying economic conditions. Such information would record reserve changes over a time
path, pin-pointed to show certain events or factors responsible for
the changes. Conceptually, it was agreed, such a breakdown would
be helpful in cost analysis, since cause and effect could be more
clearly discussed in the relations between outlay and results. Some in
the group felt that such detailed reporting would not only be exces-
sively costly, but also difficult to make internally consistent, given
the large degree of judgment which would have to be exercised by
the reporting sources. The discussion ended on the note of practica-
bility—to what extent such reporting may be feasible. This, it was
agreed, was a matter for further investigation.

In the course of the discussion, some members of the group voiced
the need for reporting reserves by reservoir as well as by geographic
region. This additional detail, they felt, could be fairly easily obtained
and would assist in the more detailed forms of cost analysis. It was

assumed that the API reporting procedures were built up from a base of reservoir information. An objection was raised that reporting on this basis would not only be difficult and costly, but that companies would refuse to cooperate in compiling such information, since it would prejudice their position in negotiating with landowners.

The matter of "dating back" reserves to the year of discovery of their fields was a thorny question. Some participants felt strongly that such action was required for determining (1) trends in additions to the physical availablity of oil and (2) trends in discovery outlay associated with this physical availability. They felt that cost analysis required additions to a known reserve to be credited back to the earlier costs of finding and developing reserves in order to get a meaningful unit cost having trend significance.

Others in the group emphasized the extreme difficulty of attempting to date back existing reserves in the absence of a suitable technique, beyond what has already been attempted by the National Petroleum Council. Although they conceded the difficulty of such reporting retrospectively, some participants insisted on the desirability of initiating this type of reporting from now forward, with respect to newly discovered fields, crediting later revisions and extensions back to the year of discovery. The consensus was that the feasibility of such reporting should be investigated.

In this matter, as in others, there sometimes appeared a difference of outlook or emphasis as between industry participants and those with an academic status. Two different kinds of tests could be applied to various proposals for improvements in data and extensions of analysis. One was whether they were significant for management in decision-making with respect to profitability. The other was whether they would serve some useful purpose in relation to questions of public policy. And in either case, are the changes worth the toil, trouble and expense involved? Everyone recognized that these questions were appropriate, so that the differences were only in the weighting of judgment.

Although the participants in the seminar were all acquainted with the conventional estimates of reserves published by the American Petroleum Institute and the American Gas Association, the discussion revealed that they felt their knowledge to be inadequate on three points: (1) the technical methods of estimating reserves, (2) the valid uses and limitations of the estimates, and (3) alternative or supplementary methods of estimating reserves which could throw additional light on future availability of petroleum. There was a general feeling that the present "official" estimates are less than sufficient to meet certain informational needs, and that, such as they are, they are commonly misinterpreted and put to invalid uses. Out of these feelings arose the suggestion that a technical study of methods of estimating reserves would serve a useful purpose. The suggestion was received with favor by Resources for the Future, Inc., which then provided a grant to support the research out of which the present study has developed.

A considerable portion of the study is devoted simply to a review of the methods used by the American Petroleum Institute and the American Gas Association in estimating reserves of crude oil, natural gas, and natural gas liquids, and to examining the uses and limitations of the estimates. Beyond that, other methods for extending the boundaries of knowledge concerning future availability of petroleum are reviewed, and various suggestions for improving the state of knowledge are examined.

The study has had the benefit of critical comment by a number of experts from both inside and outside the petroleum industry. To them we express our thanks collectively, since not all of them might care to be mentioned by name.

WALLACE F. LOVEJOY
PAUL T. HOMAN

August 1965

Contents

PART II — METHODS OF ESTIMATING RESERVES OF
NATURAL GAS AND NATURAL GAS LIQUIDS

PART III — SUMMARY AND CONCLUSION

List of Tables

Contents

xvi

Contents

8. Classification of Petroleum Reserves... 60
9. Natural Gas Reserves, 1964... 101
10. Summary of Annual Estimates of Natural Gas Reserves for Period December 31, 1945, to December 31, 1964... 102
11. Estimated Proved Recoverable Reserves of Natural Gas in the United States... 104
12. Estimated Proved Recoverable Reserves of Natural Gas Liquids in the United States... 118
13. Summary of Estimated Discoveries of Natural Gas Now Assigned to Fields Discovered in Years Shown... 123
14. Summary of Estimated Discoveries of Natural Gas Liquids Now Assigned to Fields Discovered in Years Shown... 124
15. Maximum Productive Capacity of Natural Gas in the United States. 126
16. Maximum Productive Capacity of Natural Gas Liquids in the United States... 128
17. U.S. Availability or Productive Capacity of Natural Gas Liquids, 1951-65... 128

I

METHODS OF ESTIMATING CRUDE OIL RESERVES

1

The Reserves Concept

The amounts of mineral resources available to support an expanding economy are necessarily a matter of interest to the industries which produce those resources, the industries which use them, and government agencies which are responsible for policies to promote the long-range economic well-being of the country. However, the effort to estimate the future availability of mineral resources presents difficult conceptual and technical problems. In fact, different methods of estimation for different purposes have led to considerable ambiguity and misunderstanding, as well as to efforts to mitigate the resulting confusion.[1]

In *Energy in the American Economy*,[2] the authors suggest the use of three basic concepts—reserves, resources, and resource base. Reserves, they state, "are explicitly defined in terms of immediate or short-term economic feasibility of extraction. The cost limits are consistent with normal risk taking and commercial production, and exclude material known to exist but which cannot be profitably extracted with current techniques." "Resource base" is defined as all of the particular mineral in the earth's crust within a specified geographic area, regardless of the economic or technologic feasibility of extraction. They use the term "resources" to denote that part of the resource base which can be recovered under any specified set of economic and technological conditions.

[1] For definitions and terminology, see, in particular, F. Blondel and S. G. Lasky, "Mineral Reserves and Resources," *Economic Geology*, Vol. 51, No. 7 (1956), pp. 686-97.

[2] Sam H. Schurr and Bruce C. Netschert, with Vera F. Eliasberg, Joseph Lerner, and Hans H. Landsberg, *Energy in the American Economy, 1850-1975* (Baltimore: Johns Hopkins Press for Resources for the Future, Inc., 1960).

1

These distinctions appear in petroleum industry literature but under different labels. The industry references usually use the term reserves along with some adjective or modifier to denote resources which are, may be, might be, or are not recoverable under various assumptions and conditions. Thus, as applied to the petroleum industry, the term "reserves" is subject to a number of different definitions, and the amounts under these different definitions are arrived at by different methods of estimation. When discussing reserves, it is therefore necessary to pay strict heed to the definitions and to the methods of estimation. The term must in all uses be designated with some modifying adjective or descriptive phrase to identify the definition, and must be attributed to the estimating source to identify the method of estimation.

Writing on this point in 1950, Frederic H. Lahee, an eminent petroleum geologist and former chairman of the Reserves Committee of the American Petroleum Institute (API) said:

> A very important subject is that which relates to our nation's reserves of crude oil, natural gas, and natural gas liquids. It is a subject of primary concern to the public, to the government, and to industry. It is a subject which has been widely discussed, widely misstated, and widely misunderstood. . . . when estimates of reserves of oil or of gas are published, or when references are made, verbally or in print, to these reserves, and no mention is made of the intended scope of this term "reserves," all sorts of confusion may result.[3]

For purposes of the present study, it is convenient to separate the methods of estimating crude oil reserves from the methods of estimating reserves of natural gas and natural gas liquids. The first part of the study, and the bulk of it, will be concerned with crude oil alone. This will simplify the exposition and also make it possible to deal more briefly with natural gas and natural gas liquids for which the processes of estimating reserves present analogous, but somewhat distinct, problems.

In the usage of the petroleum industry, the only aggregate figure of crude oil reserves for the United States is an estimate of "proved reserves" published annually by the American Petroleum Institute. Because it is the only "official" figure and because it is the starting point for most discussions of potential oil resources, we shall describe the estimating process in some detail in the next chapter. Proved reserves represent an estimation,

---

[3] "Our Oil and Gas Reserves: Their Meaning and Limitations," *Bulletin of the American Association of Petroleum Geologists*, Vol. 34, No. 6, June 1950.

under a rather limited and specific definition, of what may be called a "current inventory" of recoverable oil underlying existing wells within a very restricted geographic and geological circumference, or inferred to exist in immediately adjacent territory. They do not reflect the reasonable expectations of the industry concerning the amount of oil that will ultimately be recovered from known fields.

Individual companies conventionally estimate their own proved reserves, and it is largely upon the knowledge of company experts that the API estimates are built. Individual companies, in addition, extend the reserves concept to a broader use. For certain purposes they may make a separate, more speculative estimate of "probable" reserves based upon expert geological knowledge of the properties. And beyond this, they may venture into an even more speculative estimate of "possible" reserves. When members of the industry talk about reserves, their usage is usually confined to the category of proved reserves, but in some contexts they refer to a broader concept.

Outside the category of proved reserves, no agency, public or private, attempts systematically to review the evidence evaluating the amount of recoverable oil and to aggregate it into nationwide statistical summations. The public, therefore, and even the members of the industry, have no "official" statement of the expectations of the industry with respect to recoverable oil from presently known fields, according to gradations of relative certainty or uncertainty, though informal speculation on this point is common among industry experts.

The limited information on reserves supplied in the API estimates is less than adequate to satisfy the public interest in knowledge concerning future availability of crude oil. The national economy rides on the back of its energy supplies, of which petroleum is at present the major source. There is a reasonable presumption that, over the next generation or two, the domestic petroleum industry will become decreasingly capable of meeting the country's energy requirements. No responsible government can fail to interest itself in the fundamental problem of future availability of energy, in order to be forearmed with policies appropriate to the changing conditions of energy supply in relation to the mounting demand.

It appears that some of the confusion and controversy over current policy questions relating to petroleum can be attributed to the lack of understanding of reserve definitions used today and to the inability to take a longer view than these concepts allow. In the policy areas of import restriction, taxation of income from petroleum, federal leasing of offshore areas and oil shale lands, federal regulation of natural gas producers, and federal support of alternative energy supplies, there is a need to have estimates, however rough, of the quantities of reserves that can be expected under different economic and technological conditions. In response to this need, various types of studies have been made by public and private agencies, but not on a systematic basis of continuous review of the evidence.[4]

In the present study we shall be working both sides of the street. We shall be reviewing the forms of estimation customarily made within the industry and studies of the future availability of oil both within and outside the industry. The subject matter of the study is primarily the methods of estimating future availability of petroleum from presently known fields in the United States. This includes estimates of reserves in the conventional "proved" sense, and something additional about more uncertain anticipations. Some attention will also be given to views upon the possibilities of future discovery. In traversing so broad a terrain, we shall not attempt to confine the term "reserves" strictly to the limits of industry usage; but we shall attempt to avoid ambiguities by identifying different usages in the context.

Since the later text will make reference to actual figures, it is well at the outset to establish certain orders of magnitude. In doing this, we can also illustrate the difference between figures which represent reserves in the "proved" sense and those which emerge from different modes of estimating.

1. The annual production of crude oil in the United States has in recent years been running at a level not far from 2.5 billion barrels per year. Cumulative production through 1964 was roughly 75 billion barrels.

2. As of December 31, 1964, the API reported 31.0 billion barrels of crude oil "proved" reserves in presently known fields and under present economic and operating conditions (with additional "proved" reserves of natural gas liquids of 7.7 billion barrels).[5] Since the API definition of proved reserves applies very strict limiting assumptions, this is a bottom figure from which all other estimates are upward.

3. Within the industry, on the basis of experience, it is anticipated with assurance that the amount of reserves which will eventually be "proved" in known fields will be substantially larger than the amount now credited

---

[4] An effort to make periodic estimates of future natural gas supplies which go beyond estimates of proved reserves was begun in 1964 by a newly created "Future Gas Requirements and Supply Committee." This work is discussed in Chapter 11.

[5] *Proved Reserves of Crude Oil, Natural Gas Liquids and Natural Gas*, December 31, 1964, Vol. No. 19 (New York: 1965). Published jointly by American Gas Association, American Petroleum Institute, and Canadian Petroleum Association.

to those fields as proved. The additional amount is variously estimated, but an amount equal to present proved reserves roughly expresses the order of magnitude. This would mean an eventual recovery of possibly 60 billion barrels from known fields without striking changes in operating or economic conditions, but with considerable expansion of secondary recovery activity.

4. Substantial additions to reserves are anticipated from the application of improved recovery techniques, though the amount of such additions cannot be closely estimated. A very modest attribution of 10 billion barrels to this source would bring recoverable "reserves" up to 70 billion barrels. If the recovery rate could be increased to 50 per cent of oil originally in place (346 billion barrels as estimated by a committee of the Interstate Oil Compact Commission[6]) the amount would rise to over 100 billion barrels (after allowing for oil already produced). Such figures become highly speculative, and allow play for different degrees of optimism and pessimism, but even the largest figure cited above lies within what some well-informed investigators consider a reasonable range of possibilities.

5. Further future additions will come from new discoveries outside the producing horizons of presently known fields, the results of which are highly speculative. Apart from the uncertain incidence of oil even in favorable geological environments, much will depend upon the strength of the economic incentives to exploratory effort and upon improvements in finding technology. Some recent attempts to quantify these potentialities will be noted at later points.

On the preceding showing, it is clear that anyone who wishes to penetrate the world of oil reserves estimates must inform himself of what the various sets of figures mean. On this account, we shall review a number of studies which estimate, or propose methods of estimating, oil "reserves," using the word in a loose sense. If one wishes to stick to the strict industry usage, much of what we will have to say is not about "reserves," but about something which must be described as "estimates of the future availability of oil resources under various assumptions." It is verbally easier to use the word "reserves" with the understanding that it means different things in various defined contexts.

---

[6] Paul D. Torrey, "Evaluation of United States Oil Reserves as of January 1, 1962," *Oil and Gas Compact Bulletin*, Vol. XXI, No. 1, June 1962.

2

The API Estimating Procedures

Since the American Petroleum Institute (API) estimates of "proved" reserves are the most widely used and provide the point of departure for most other estimates, we now turn to examine in some detail the procedures used in making these estimates.

Each year since 1936, the API has estimated and published "oil" reserve data for the United States, by states.[1] From 1936 through 1945 the "oil" reserves included crude oil reserves plus cycle-plant and lease condensate.[2] In 1945, the condensate was estimated to be about $4\frac{1}{4}$ per cent of the total "oil" reserve. Clearly, much liquid occurring with natural gas never got into these figures.

In 1946, the definitions were changed so that "crude oil" and "natural gas liquids" (NGL) were reported separately, and the two were combined to give an estimate of total "liquid hydrocarbon" reserves. In 1947, almost 13 per cent of the total liquid hydrocarbon reserve was attributed to natural gas liquids, a substantially larger proportion than when only cycle-plant and lease condensate was included. The estimates for NGL reserves were made in conjunction with the American Gas Association, which began reporting natural gas reserves in 1946. Thus, there is a series of liquid hydrocarbon reserve figures, based on uniform definitions, reported annually and broken down between crude oil and NGL from 1946 to the present.

[1] *Proved Reserves of Crude Oil, Natural Gas Liquids and Natural Gas*, December 31, 1964, Vol. No. 19 (New York: 1965). Published jointly by American Gas Association, American Petroleum Institute, and Canadian Petroleum Association.

[2] Lease condensate is condensate produced in conjunction with crude oil, usually from wells in which they are joint products.

6

The machinery through which API reserve figures are collected and aggregated is a rather complex, self-directed system of committees and subcommittees. The parent committee is the API Committee on Petroleum Reserves, currently composed of thirteen men, eleven from different major oil companies, one from a state geological survey, and one from the API staff. Most of the committee members have reporting to them one or more of twenty or thirty subcommittees which are dispersed geographically in the major oil-producing regions of the nation. There were in 1964 about 140 subcommittee members, men who are technically competent and who are familiar with the fields to be covered by their particular subcommittees.[3]

Since detailed reserves data are usually considered to be highly confidential by the companies owning the reserves, a subcommittee member reporting on a particular reservoir must rely (1) on his own company's information; (2) on information he can glean from scouts, landmen, lease brokers, and so forth; and (3) on information he receives in the strictest confidence from other operators in the field. In cases in which a new field is discovered or in which a field is still in the early stages of development, leasing can be extremely competitive and reserves information highly secret. In such instances, the subcommittee member must rely primarily on his own sources of information, and initial estimates may involve some guesswork. In cases where a field has been developed and further lease play is unlikely, he can usually get information on the geology of the fields, well logs, core sample reports, and production and reservoir performance records, so that estimates of reserves can be made with considerably more accuracy. The subcommittees are so constituted and the areas covered by different members are so assigned that the data reported can be aggregated to show the reserves in each producing state as well as in the country at large. Much information is in the hands of the state conservation agencies, but they do not officially participate in the estimating procedures.

API subcommittees meet in January and February to compile the reserves data for their assigned areas. Each member brings in his reports. In most instances the figures reported by a subcommittee member responsible for a certain district are accepted as the best possible, since there is often no way to check them. In the case of large fields, subcommittee discussion usually precedes the setting of a figure for that year. The subcommittee chairmen, who are selected with great care by the parent API reserves committee, compile on worksheets the data for each assigned area.

---

[3] For a lucid, short account of the method and philosophy of the estimates, see an article by Morris Muskat, chairman of the API Committee, "The Proved Crude Oil Reserves of the U. S.," *Journal of Petroleum Technology*, September 1963.

These worksheets are kept by the subcommittee chairman, though they may be shown at meetings of the parent committee. Thus, the parent committee has no detailed records of reserves data. Procedures vary among subcommittees, so no accurate general statement can be made of the relation between subcommittees and the parent committee. Subcommittee chairmen are impressed by the parent committee with the need for maintaining consistent definitions and concepts.

The parent reserves committee meets for only a few days each year, at which time the committee members submit the aggregated figures for the area, or areas, under their general supervision. Committee discussion usually relates only to major fields or to some particularly difficult or controversial aspect. Differences in opinion are usually compromised. Each year some subcommittee chairmen are invited to sit with the parent committee to gain an appreciation of how the committee works and what some of its problems are. The turnover of membership on both the parent committee and the subcommittees is slow so that a substantial degree of continuity is achieved from year to year. There have been only three chairmen of the API parent committee since its inception in 1936.

THE API QUANTITATIVE ESTIMATES

For purposes of later reference, we start by reproducing the latest summary API tables which give the estimates of proved reserves. The annual summary for 1964 is shown in Table 1. Table 1A shows crude oil alone, Table 1B natural gas liquids alone, and Table 1C the two combined. Since liquids derived from natural gas go into many of the same final products as crude oil does, the reserves are a significant part of the total. But since estimating them is connected with estimates of natural gas reserves, the problems of estimation will not be taken up in the present part of this paper, which considers crude oil only.

The time series prepared from these annual estimates is presented in Table 2, starting from 1946 when new definitions were imposed. The table shows crude oil alone, natural gas liquids alone, and the two combined.

Table 3 provides a breakdown by states of the estimates of reserves of crude oil for 1964. In the original source, natural gas liquids and the combined figures are also given by states.

TABLE 1. PROVED RESERVES OF CRUDE OIL, NATURAL GAS LIQUIDS, AND TOTAL LIQUID HYDROCARBONS, 1963

(Thousands of barrels of 42 U.S. gallons)

A. *Proved Reserves of Crude Oil*

Total proved reserves of crude oil as of December 31, 1963............		30,969,990
Revisions of previous estimates....................... (+)	899,292	
Extensions of old pools............................	1,419,182	
New reserves discovered in 1964 in new fields and in new pools in old fields...............................	346,293	
Proved reserves added in 1964...................................		2,664,767
Total proved reserves as of December 31, 1963 plus new proved reserves added in 1964..		33,634,757
Less production during 1964.......................................		2,644,247 [a]
Total proved reserves of crude oil as of December 31, 1964............		30,990,510
Change in crude oil reserves during 1964..........................		(+) 20,520

B. *Proved Reserves of Natural Gas Liquids*

Total proved reserves of natural gas liquids as of December 31, 1963....		7,673,978
Revisions of previous estimates and extensions of old pools.. (+)	457,702	
New reserves discovered in 1964 in new fields and in new pools in old fields...............................	151,042	
Proved reserves added in 1964...................................		608,744
Total proved reserves as of December 31, 1963 plus new proved reserves added in 1964..		8,282,722
Less production during 1964.......................................		536,090 [a]
Total proved reserves of natural gas liquids as of December 31, 1964....		7,746,632
Change in natural gas liquids reserves during 1964..................		(+) 72,654

C. *Proved Reserves of Total Liquid Hydrocarbons*

(A and B combined)

Total proved reserves as of December 31, 1963......................		38,643,968
Revisions of previous estimates and extensions of old pools (+)	2,776,176	
New reserves discovered in 1964 in new fields and in new pools in old fields...............................	497,335	
Proved reserves added in 1964...................................		3,273,511
Total proved reserves as of December 31, 1963 plus new proved reserves added in 1964..		41,917,479
Less production during 1964.......................................		3,180,337 [a]
Total proved reserves of liquid hydrocarbons as of December 31, 1964..		38,737,142
Change in total liquid hydrocarbons reserves during 1964.............		(+) 93,174

[a] Production figures in part estimated.

SOURCE: *Proved Reserves of Crude Oil, Natural Gas Liquids, and Natural Gas in the United States and Canada*, December 31, 1964, Vol. No. 19 (New York: 1965). Published jointly by the American Gas Association, American Petroleum Institute, and Canadian Petroleum Association.

Estimating Crude Oil Reserves

TABLE 2. SUMMARY OF PROVED RESERVES AS REPORTED FOR 1946 AND THEREAFTER [a]

(Thousands of barrels of 42 U.S. gallons)

| | New Oil Added during Year | | | | | |
Year	Revisions of previous estimates and extensions to known fields (1)	Discoveries of new fields and new pools in old fields (2)	Total new discoveries, extensions, and revisions (columns 1 + 2) (3)	Production during year [b] (4)	Estimated proved reserves as of end of year (5)	Increase over previous year (column 3 − 4) (6)
			Crude Oil Only			
1946	2,413,628	244,434	2,658,062	1,726,348	20,873,560	931,714
1947	2,019,140	445,430	2,464,570	1,850,445	21,487,685	614,125
1948	3,398,726	396,481	3,795,207	2,002,448	23,280,444	1,792,759
1949	2,297,428	890,417	3,187,845	1,818,800	24,649,489	1,369,045
1950	1,997,769	564,916	2,562,685	1,943,776	25,268,398	618,909
1951	4,024,698	389,256	4,413,954	2,214,321	27,468,031	2,199,633
1952	2,252,860	496,428	2,749,288	2,256,765	27,960,554	492,523
1953	2,704,450	591,680	3,296,130	2,311,856	28,944,828	984,274
1954	2,287,231	585,806	2,873,037	2,257,119	29,560,746	615,918
1955	2,393,767	476,957	2,870,724	2,419,300	30,012,170	451,424
1956	2,507,114	467,222	2,974,336	2,551,857	30,434,649	422,479
1957	2,008,603	416,197	2,424,800	2,559,044	30,300,405	(−)134,244
1958	2,293,513	314,729	2,608,242	2,372,730	30,535,917	235,512
1959	3,297,383	369,362	3,666,745	2,483,315	31,719,347	1,183,430
1960	2,111,472	253,856	2,365,328	2,471,464	31,613,211	(−)106,136
1961	2,296,193	361,374	2,657,567	2,512,273	31,758,505	145,294
1962	1,800,310	380,586	2,180,896	2,550,178	31,389,223	(−)369,282
1963	1,824,219	349,891	2,174,110	2,593,343	30,969,990	(−)419,233
1964	2,318,474	346,293	2,664,767	2,644,247	30,990,510	20,520
			Natural Gas Liquids Only			
1946	(......This detail not available for 1946......)				3,163,219	....
1947	192,237	59,301	251,538	160,782	3,253,975	90,756
1948	405,874	64,683	470,557	183,749	3,540,783	286,808
1949	294,211	92,565	386,776	198,547	3,729,012	188,229
1950	707,879	58,183	766,062	227,411	4,267,663	538,651
1951	648,497	75,494	723,991	267,052	4,724,602	456,939
1952	475,170	81,668	556,838	284,789	4,996,651	272,049
1953	648,047	95,922	743,969	302,698	5,437,922	441,271
1954	20,830	86,520	107,350	300,815	5,244,457	(−)193,465
1955	447,160	67,348	514,508	320,400	5,438,565	194,108
1956	715,764	94,056	809,820	346,053	5,902,332	463,767
1957	8,884	128,508	137,392	352,364	5,687,360	(−)214,972
1958	749,956	108,250	858,206	341,548	6,204,018	516,658
1959	593,905	109,539	703,444	385,154	6,522,308	318,290
1960	603,621	121,509	725,130	431,379	6,816,059	293,751
1961	590,537	104,149	694,686	461,649	7,049,096	233,037
1962	580,570	151,979	732,549	470,128	7,311,517	262,421
1963	700,183	177,937	878,120	515,659	7,673,978	362,461
1964	457,702	151,042	608,744	536,090	7,746,632	72,654

TABLE 2. SUMMARY OF PROVED RESERVES AS REPORTED FOR 1946 AND THEREAFTER ᵃ—Continued

(Thousands of barrels of 42 U. S. gallons)

Year	New Oil Added during Year			Production during year ᵇ (4)	Estimated proved reserves as of end of year (5)	Increase over previous year (column 3 − 4) (6)
	Revisions of previous estimates and extensions to known fields (1)	Discoveries of new fields and new pools in old fields (2)	Total new discoveries, extensions, and revisions (columns 1 + 2) (3)			
	Total Liquid Hydrocarbons					
1946	(......This detail not available for 1946......)				24,036,779	....
1947	2,211,377	504,731	2,716,108	2,011,227	24,741,660	704,881
1948	3,804,600	461,164	4,265,764	2,186,197	26,821,227	2,079,567
1949	2,591,639	982,982	3,574,621	2,017,347	28,378,501	1,557,274
1950	2,705,648	623,099	3,328,747	2,171,187	29,536,061	1,157,560
1951	4,673,195	464,750	5,137,945	2,481,373	32,192,633	2,656,572
1952	2,728,030	578,096	3,306,126	2,541,554	32,957,205	764,572
1953	3,352,497	687,602	4,040,099	2,614,554	34,382,750	1,425,545
1954	2,308,061	672,326	2,980,387	2,557,934	34,805,203	422,453
1955	2,840,927	544,305	3,385,232	2,739,700	35,450,735	645,532
1956	3,222,878	561,278	3,784,156	2,897,910	36,336,981	886,246
1957	2,017,487	544,705	2,562,192	2,911,408	35,987,765	(−)349,216
1958	3,043,469	422,979	3,466,448	2,714,278	36,739,935	752,170
1959	3,891,288	478,901	4,370,189	2,868,469	38,241,655	1,501,720
1960	2,715,093	375,365	3,090,458	2,902,843	38,429,270	187,615
1961	2,886,730	465,523	3,352,253	2,973,922	38,807,601	378,331
1962	2,380,880	532,565	2,913,445	3,020,306	38,700,740	(−)106,861
1963	2,524,402	527,828	3,052,230	3,109,002	38,643,968	(−) 56,772
1964	2,776,176	497,335	3,273,511	3,180,337	38,737,142	93,174

ᵃ Comparable data for earlier years not available.
ᵇ Data on production partially estimated. Cumulative production not additive.
SOURCE: *Proved Reserves of Crude Oil, Natural Gas Liquids, and Natural Gas in the United States and Canada*, December 31, 1964, Vol. No. 19 (New York: 1965). Published jointly by the American Gas Association, American Petroleum Institute, and Canadian Petroleum Association.

TABLE 3. ESTIMATED PROVED RESERVES OF CRUDE OIL IN THE UNITED STATES (API COMMITTEE) [a]

(Thousands of barrels of 42 U. S. gallons)

State	Proved reserves as of December 31, 1963 (1)	Changes in proved reserves due to extensions and revisions during 1964 (2)	Proved reserves discovered in new fields and in new pools in old fields in 1964 [b] (3)	Pro-duction during 1964 [c] (4)	Proved reserves as of December 31, 1964 (columns 1 + 2 + 3 less column 4) (5)	Changes in reserves during 1964 (column 5 less column 1) (6)
Alaska	74,920	19,000	—	11,054	82,866	7,946
Alabama	45,069	13,537	—	8,710	49,896	4,827
Arkansas	225,291	5,191	615	25,937	205,160	(−) 20,131
California [d]	3,599,735	785,347	39,275	298,876	4,125,481	525,746
Colorado	368,375	9,775	2,585	34,501	346,334	(−) 22,041
Illinois	416,612	44,859	890	70,825	391,536	(−) 25,076
Indiana	63,432	8,135	164	11,102	60,629	(−) 2,803
Kansas	841,349	51,024	10,034	105,866	796,541	(−) 44,808
Kentucky	100,456	32,737	5,005	20,337	117,861	17,405
Louisiana [d]	5,088,605	385,777	166,564	478,458	5,162,488	73,883
Michigan	68,543	5,068	185	15,497	58,299	(−) 10,244
Mississippi	384,909	22,011	5,451	55,804	356,567	(−) 28,342
Montana	271,253	4,745	6,290	30,668	251,620	(−) 19,633
Nebraska	83,583	2,816	3,450	18,748	71,101	(−) 12,482
New Mexico	1,010,729	48,251	6,148	108,336	956,792	(−) 53,937
New York	18,435	(−) 3,000	—	1,874	13,561	(−) 4,874
North Dakota	389,158	9,763	3,914	25,599	377,236	(−) 11,922
Ohio	87,814	27,598	—	15,850	99,562	11,748
Oklahoma	1,628,138	149,925	6,701	198,879	1,585,885	(−) 42,253
Pennsylvania	91,734	—	—	5,113	86,621	(−) 5,113
Texas [d]	14,573,125	585,109	70,309	928,696	14,299,847	(−)273,278
Utah	219,576	25,017	3,315	28,400	219,508	(−) 68
West Virginia	57,303	4,693	—	3,286	58,710	1,407
Wyoming	1,254,306	77,939	12,960	140,669	1,204,536	(−) 49,770
Miscellaneous [e]	7,540	3,157	2,438	1,262	11,873	4,333
Total United States	30,969,990	2,318,474	346,293	2,644,247	30,990,510	20,520

[a] Includes lease condensate produced with crude oil.

[b] Only a limited area is assigned to each new discovery, even though the committee may believe that eventually a much larger area will produce; for, in this report, the concern is only with actually proved reserves.

[c] Data on production partially estimated.

[d] Includes off-shore reserves.

[e] Under Miscellaneous are included Arizona, Florida, Missouri, Nevada, South Dakota, Tennessee, and Virginia.

SOURCE: *Proved Reserves of Crude Oil, Natural Gas Liquids, and Natural Gas in the United States and Canada*, December 31, 1964, Vol. No. 19 (New York: 1965). Published jointly by the American Gas Association, American Petroleum Institute, and Canadian Petroleum Association.

THE DEFINITION OF PROVED RESERVES

The parent committee establishes the criteria for estimating "proved" reserves. Insofar as it is able, it attempts to induce uniformity in the application of these criteria by the numerous subcommittee members who do the actual estimating. There is some leeway for individual or regional interpretation of the definitions and working rules; but whatever regional biases or differences exist, they are likely to remain relatively constant because of the extremely slow turnover of subcommittee membership.

The basic definition and working rules for estimating crude oil proved reserves are as follows:

The reserves listed in this Report, as in all previous Annual Reports, refer solely to "proved" reserves. These are the volumes of crude oil which geological and engineering information indicate, beyond reasonable doubt, to be recoverable in the future from an oil reservoir under existing economic and operating conditions. They represent strictly technical judgments, and are not knowingly influenced by policies of conservatism or optimism. They are limited only by the definition of the term "proved." They do not include what are commonly referred to as "probable" or "possible" reserves.

The proved reserves may be considered as the known and established underground working inventory available for recovery under prevailing conditions. These estimates are subject to future revisions, either downward or upward, even though the *presently established* "proved" reserves may be accurate, in the light of current information.

Both drilled and undrilled acreage are considered in the estimates of the proved reserves. However, the undrilled proved reserves are limited to those drilling units immediately adjacent to the developed areas which are virtually certain of productive development, except where the geological information on the producing horizons insures continuity across the undrilled acreage.

The proved crude oil reserves estimates do not include:

1) Oil whose recovery is subject to reasonable doubt because of uncertainty as to geological conditions, reservoir characteristics or economic factors.

2) Oil in untested prospects.

3) Oil that may become available by fluid injection or other methods from fields in which such operations have not yet been applied.

4) Liquid hydrocarbons that may become available through the processing of natural gas.

5) Oil that may be recovered from oil shales, coal or other substitute sources.

The reserves reported here for the various states are the totals of those for all the individual fields and reservoirs in these states which are still capable of economic production at the year end. Each year initial estimates are made for the new pool and the new field dis-

coveries. The earlier fields are reviewed annually to adjust the estimates of the previous year for the effects of the year's production,
extensions of the proved area, if any, and of more completely defined
performance, reservoir and economic factors which may indicate
the need for revisions in the recovery estimates. The composite balance among these basic elements gives the new year-end reserves
estimates.[4]

The critical words in this definition are "beyond reasonable doubt" and
"under existing economic and operating conditions." They provide the
basis for the character of the estimates.

• "Beyond reasonable doubt" is a principle that largely eliminates the
whole realm of degrees of certainty. In an industry in which uncertainty
is the prime fact of life, the elimination of doubt necessarily excludes much
that informed men "know" to exist. Every informed person knows that,
"beyond reasonable doubt," the ultimate recovery from the aggregate of
presently known fields will be substantially larger than the amount stated
in the official current estimates of proved reserves. This knowledge is based
upon historical and statistical information. It is also logically inherent in
several aspects of the process of estimation itself.

1. In the first place, the figures for new discoveries in any year are
limited by the specified, restrictive definitions. Very little oil is *proved*
recoverable in the year of discovery for any given field or well. But the
statistical evidence shows that the initial amounts will, in the aggregate,
be multiplied several times over by extensions and revisions in later years.

2. The defined basis of estimating extensions due to early development
drilling also leaves an unestimated accrual to be added in later years.

3. The same may be said of additions in the field of revisions based on
increasing knowledge of the geological characteristics of developed fields.

4. Later additions will also accrue from new fluid injection (secondary
recovery) projects, when pilot plants or installed operations provide the
basis for an estimate of increased recovery.

The strict official definition of "proved" excludes any attempt at quantitative estimation of these additional amounts, into which a large element
of uncertainty necessarily enters. Nevertheless, experience shows in a rough
way how initial discovery estimates are enlarged by later extensions and
revisions. For the fifteen years, 1948–62, the extensions and revisions for

[4] The quotations found in this section are from API-AGA-CPA, *Proved Reserves of
Crude Oil, Natural Gas Liquids, and Natural Gas in the United States and Canada,*
December 31, 1964, *op. cit.*

crude oil averaged 5.2 times the proved reserves from new discoveries. By five-year periods, the averages were 4.5 in 1949–53, 5.0 in 1954–58, and 6.6 in 1959–63. This would imply that in general a factor of 5 to 6 could be applied to original discoveries as an estimate of future extensions and revisions, but there is no assurance about the future size of the multiplier.

• "Under existing operating conditions" is a technical principle which limits the size of reserves estimates. For example, it excludes all the secondary recovery opportunities which are possible, but not yet introduced, even under "existing economic conditions." According to a study sponsored by the Interstate Oil Compact Commission,[5] admitting this factor would increase even strictly defined "reserves" by about 50 per cent. This exclusion of economical secondary recovery possibilities emphasizes the reliance of the API estimates upon "facts"—facts in the sense of the actual existing operating conditions, as against the facts of known methods of economical recovery. "Economical recovery" in this context becomes a rather ambiguous concept. If companies had no other uses for their funds—if, for example, it were known that funds spent on exploration would not yield remunerative results—it would "pay" to expand secondary recovery operations. In practice, with the limited funds available, it is necessary to make decisions as to the amounts to be expended in different directions. These decisions exclude what would, in other circumstances, be economical expenditures on secondary recovery.

"Under existing operating conditions" also ignores reserves which might be "proved" if the most modern and economical existing technology were applied more extensively and more intensively. Industry literature is replete with examples of "good" and "bad" practices in drilling, completion, and production. The API definition takes operating conditions as they are, good and bad, and makes no judgments concerning reserves which might be included in the proved category from greater application of proven techniques.

The specification of "under existing operating conditions" also excludes any consideration of the possible results of improved technological methods of recovery. Every informed member or student of the industry expects these improvements to occur, and they reasonably enter into any discussion of probability. With its emphasis on "facts," the API is possibly wise in excluding the possibility of improved technology, which is notoriously difficult to forecast. It is not a measurable factor. But in doing so,

[5] Paul D. Torrey, "Evaluation of United States Oil Reserves as of January 1, 1962," *Oil and Gas Compact Bulletin*, Vol. XXI, No. 1, June 1962.

the API estimates downgrade the expectations of the ultimate recovery which informed persons expect within various degrees of probability.

• "Under existing economic conditions" stresses a distinction, which has to be made, between physically recoverable oil and oil for which the incentives exist to recover it. The economic conditions, although not specifically spelled out by the API, must include such things as the price and production of oil (which together yield gross revenue), costs of performing the producing function, prices of competitive energy sources, costs of finding and developing new petroleum supplies, expected or target rates of profit, and existing federal, state, and local tax structures. Every oil well has a life of generally declining productivity (which may be modified but not basically changed by conservation measures and the introduction of pressure maintenance or secondary recovery projects). At some point in time, as daily production gets smaller and smaller, it becomes "uneconomical" to continue to operate the well, that is, the out-of-pocket costs of continuing operation catch up with the revenues from such an operation. Clearly, this is an economic decision that each company must make for each well it owns. The abandonment point shifts in the time dimension with changes in the price of the product, with changes in out-of-pocket costs, with changes in taxes, with changes in interest rates and the profitability of alternative investments. If the abandonment point moves closer because of such things as higher costs or lower prices, then reserves must be revised downward. If this point is pushed further into the future by such things as lower costs or higher prices, reserves are augmented.

While in principle every change in economic conditions should have an effect on the estimation of reserves, it is probable that they do not have a very marked effect on the year-to-year API estimates. Relatively few wells or pools are truly at the economic margin in any given year, so that even a fairly drastic change in economic conditions would not affect the total national figures very much (although the impact on some specific states might be significant). Since exploration and development costs are "sunk," companies will produce with an eye to revenues in excess of out-of-pocket producing costs. These are often quite low relative to revenues until production gets quite low itself. Naturally, any marked trend in costs and prices would affect the incentives for exploratory and developmental drilling, and thus affect the course of future estimates of reserves.

Contrary to orthodox reservoir engineering thinking in earlier years, it is now held that marginal wells can, in many cases, be stopped and started without any great effect upon ultimate recovery. What this means is that the reserves in marginal fields may go in and out of the reserve totals, depending on economic conditions. In some cases wells may be drowned out by intermittent production; and well maintenance in other cases may be costly; but very often the reservoir itself may be little affected.

Enough has been said to show that the definition of proved reserves, however appropriate to its purpose of quantifying a working inventory, is in no way a guide to reasonable expectations of the amount of oil to be recovered from presently known fields.

THE CATEGORIES OF CHANGES IN RESERVES

Gross reserves of crude oil added each year in the API estimates are broken down into two broad categories: those attributable to "discoveries" and those attributable to "extensions and revisions." For the national aggregate, "extensions" and "revisions" are shown separately, but for the individual states they are lumped together. The essential distinction between the two terms is that "extensions" represent a geographical extension of the producing territory of known fields, or reservoirs within them, while "revisions" reflect increased knowledge of the producing potentialities of existing fields or the application of improved recovery methods. Thus, extensions reflect primarily increases in the "acreage factor" while revisions reflect primarily increases in the "recovery factor." The distinction, however, often becomes quite hazy.

• "Discoveries," according to the API Committee definition and discussion of reserves, are explained in this way:

Newly discovered petroleum reservoirs, even in existing fields, seldom are fully developed during the year of discovery. Therefore, the year-end reserves estimates of discoveries generally represent only a part of the reserves that ultimately will be assigned to these new reservoirs. Where drilling beyond the discovery well has not occurred during the year, the proved area assigned to the reservoir is usually small. For this reason, reserves from discoveries are generally a comparatively small part of the total reserves additions during the year.

Reserves attributed to discoveries are those reserves found by exploratory drilling in new pools in new fields or new pools in old fields. The initial estimate of a discovery is invariably quite small, with respect to both the acreage factor (areal extent) and the recovery factor (ability to get out the petroleum in place). The acreage allowed for a discovery well will rarely include initially more than the discovery location plus the offset drilling locations[6] dictated by known geology—for example, perhaps only the indicated yield from an area of 40 acres, or perhaps for an area of 200 acres in a field with 40-acre spacing and attractive geology. This may be true even though the over-all geological and geophysical evidence indicates a reservoir of considerably greater areal extent. A pool discovered late in the year with only one well drilled would show a smaller "discovery" than one discovered earlier in the year with supplementary drilling. The recovery factor is apt to be understated initially because the information available is limited; a solution gas drive is usually assumed; and a natural water drive is rarely anticipated in early estimates.

The category of discoveries includes both "new fields" (or "rank wildcats") and "new pools in old fields." The latter designation may include either the results of "outpost" drilling within or along the edge of existing fields or pools above or below an existing producing pool, and may even result from the deepening of existing producing wells.

- "Extensions" as a category are described by the API as follows:

Extensions of existing reservoirs normally provide important new reserves additions each year. A discovery during one year will usually result in the drilling of additional wells during subsequent years. Generally these wells add productive acreage to the previously estimated proved area. Development drilling continues to generate reserves extensions until the field or reservoir is fully developed.

"Extensions" is primarily a geographical category. As the offset locations to a discovery well are drilled and found productive, then extensions to the initial discovery estimates may be further added by the offset locations to these first offset wells. The reserves added in this instance would fall into the extension category. It is quite obvious that reserves are "proved up" only slowly, and only by drilling in the case of extensions. It will be recalled that "proved" reserves include both "drilled and undrilled acreage," but the latter only when the "drilling units immediately adjacent to the developed areas are virtually certain of productive development."

---

[6] An offset location is a drilling site immediately adjacent to a producing well; for example, a discovery well drilled on 40-acre spacing may have four 40-acre offset locations, one each on the east, west, north, and south.

Since later drilling may prove some of the earlier expectations in this respect to be unjustified, in theory, at least, there may be "negative extensions," though these are officially classified as "revisions."

- "Revisions" is a category designated by the API as follows:

Additional producing wells in a reservoir not only add to the estimated productive area but they usually also add to and improve upon the basic geological and engineering data. Early estimates of porosity, interstitial water, pay thickness and other important reservoir factors may, therefore, be revised by the additional data from later wells. Also, as field development continues, production history accumulates and more accurate concepts of the reservoir performance, producing mechanism and recovery efficiency are formulated. The composite of this new and improved information will result in more precise estimates of the ultimate recoveries and remaining reserves. The changes from earlier estimates, either upward or downward, arising from such reevaluations or the installation of processes for increasing recovery constitute the revisions exhibited in the overall reserves balance. These may represent a significant part of the total changes from year to year. A minor part of the revisions total is generally provided by corrections of the estimates of the previous year's production, as finalized during the current year.

As the above description suggests, reserves added by way of revisions will commonly raise the anticipated recovery factor and can be the result of any of a number of things: improved knowledge of the geological characteristics of the field and introduction of new methods of recovery and reservoir management, such as gas reinjection, water flooding, or other methods of pressure maintenance and secondary recovery. As improvements in recovery technology are gradually introduced, they make their way into the reserves figures. Secondary reserves are usually not credited to the reservoir until after the pilot project stage has been proved successful. Also it is through revisions that changing economic conditions are reflected in reserves. Reserves may, on occasion, be reduced through revisions when increased knowledge of a field lowers the recovery factor or reduces the areal extent. While it would be a matter of great interest, if available, the data as reported provide no way of separating out the additions to, or reductions from, reserves due to improved information on the one hand and to application of secondary recovery or improved technology on the other.

SPECIAL REPORTING PROBLEMS

While the categories of discovery, extension and revision are clear enough in principle, the attribution of added reserves to these categories presents problems which can at times only be resolved by arbitrary designation. For example, how far does an outpost have to be from a known field to constitute a discovery well? How should a well which hits a new pool in a known producing horizon be treated? And should deeper pay zones completed through old wells be classified as discoveries or extensions? If an offset well proves up new acreage but yields information leading to a revision in the recovery factor, how should the changes be cataloged as between extensions and revisions?[7] These are but a few of the questions the API subcommittees have to answer. Such questions are inherent in the nature of the estimating process and cannot be avoided. As we shall see, however, there remains a general question for the API Committee, whether it might not break down the data in ways which would show quantitatively the reasons for changes in reserves, especially in the category of revisions.

One anomaly in the reserve reporting calls for attention. In the reporting of crude oil reserve data, the API separates extensions from revisions for the nation as a whole, as is shown in Table 1, for 1964. This, however, is not done for the state totals, where the two categories are reported only in combined form. The apparent reason for the failure to report the revision-extension breakdown by states is a fear of revealing confidential company data in some states where one or two large producers dominate the field. Since the national total must be built up field by field and state by state, a state breakdown between revisions and extensions could be made, with a lumping together of states in which confidential data might be revealed. The separation of state data into revisions and extensions would be helpful in pinpointing geographically trends in discovery, development and secondary recovery.

In the case of natural gas liquids and natural gas, extensions and revisions are reported in combined form for the nation as well as for the states. It seems likely here also that the aggregate must be built up from data on extensions by fields and by states, and from revisions by fields and

---

[7] A classification of exploratory wells has been developed but never related to the reserves found by each class of wells. See J. Ben Carsey and M. S. Roberts, "Exploratory Drilling in 1962," *Bulletin of the American Association of Petroleum Geologists*, Vol. 47, No. 6 (June 1963), p. 891.

by states. The major stumbling block may be a lack of consistency on the part of the subcommittees in gathering and reporting these types of data. Certainly the definitional problems are formidable.

A reporting problem faced by the analyst who uses API reserves data and the specified definitions is the omission, in the reports, of any discussion of the statistical reliability of the series. The preceding discussion has indicated that any estimate of unproduced oil involves some degree of uncertainty, even within the specified definitions. It would appear possible, based on historical experience, to provide reasonable ranges of possible error in the three categories of reported reserves.

THE ESTIMATING PERSONNEL

It was mentioned earlier that company representation on the API parent reserves committee is from major companies only. This is largely true on the subcommittees as well, although it is not exclusively so.[8] The reason why major company representation dominates the estimating committees is presumably that only the expert knowledge contained within these companies makes the estimating process possible. Every major oil company has estimates of its own reserves and in some cases of all proved reserves in areas in which it operates or is considering entry, together with additional information concerning the production prospects. At least one company, and possibly one or two others, keeps reserves data on all fields in the United States and Canada. Others do the same on a more restricted regional basis. Much care and expense goes into these company estimates since their own reserve figures provide the basis for accounting and tax reporting, and their wider knowledge of reserves is part of the information upon which investment decisions are based.

Companies commonly use a concept of "proved" reserves; but companies estimating their own reserves for internal company purposes need not follow the exact definition of the API. Most companies probably follow the API definition rather closely, but in some instances company estimates of "discoveries" tend to be substantially larger, because they give more weight to their expectations based on geological and geophysical information.

Companies keep reserves data in considerable detail—by fields, by

---

[8] Of the 143 subcommittee members serving in 1964, 109 were employees of what might be called major integrated companies. Eighteen of the remaining 34 served on the subcommittees for Tennessee, Ohio, West Virginia, New York, and Pennsylvania.

reservoirs within fields, and often by lease. In addition, they sometimes make analyses in some detail on the causes of revisions and extensions. Obviously, also, they cannot be concerned solely with "proved" reserves, however defined. For planning purposes and for investment decisions, they have to consider orders of probability. Each year or half-year, some companies review each producing and prospective property to determine how much, if any, money should be put into it, or whether the property should be abandoned. Since investment opportunities are always far more plentiful than investment funds, this periodic review serves as a rationing device which allocates scarce funds among alternative uses with an eye toward maximizing long-run over-all returns.

The bearing of all this upon the personnel and procedures of the API estimating process is that the twenty or so companies which do detailed reserves analysis for their own internal purposes possess a wealth of expert talent and technical information which can be tapped by the API Committee. It is not surprising, therefore, that the major companies provide most of the personnel, since they have the staff and funds required for sophisticated reserves studies for internal company use. This talent and information, brought to focus in the men who make up the API subcommittees, is what makes possible the API estimates without imposing a backbreaking task of reserves data collection.

This brings up another relevant point. A glance at the make-up of the subcommittees reveals the obvious fact that the members do not have complete information on all fields. Much production is in the hands of small, nonintegrated producing companies. While the major companies, among themselves, are likely to have fairly comprehensive information on most fields, to some minor extent the subcommittees may have to do some guessing about fields for which the available information is less complete. Expanding the API subcommittees in order to overcome this deficiency would undoubtedly make them more unwieldy, and would probably add little information.

Another point is to be noted. Some of the reserves information possessed by companies is considered highly confidential and has its value to them as the basis for lease acquisitions and development planning. To a considerable degree, the subcommittee members are released from the necessity of providing such information by the restricted API definition of "proved" reserves. By confining the additions from "discoveries" to a very limited areal extent, no disclosure is required from companies as to their expectations concerning the new properties. Similarly, the manner of reporting the results of extensions and new information provides substantial protection against disclosure of significant confidential information. Where even within the context of API definitions and procedures, important and confidential information would be called for, the problem can be met either by failure to report, or by masking the information in such a way that its company origin is not disclosed, or by spreading the addition or reduction over several years. We gain the impression that competitive concealment has a rather minor effect upon the estimates.

To put these matters briefly, the reporting of reserves takes place in an environment of competitive relations among companies. Confidential reserves information is an instrument through which companies hope to gain an advantage over their competition. Consequently, any plan of reporting which is sponsored by the industry must be such as to protect the members of the industry from the necessity of reporting confidential information. This presumably is one reason why the API estimates of reserves lie almost wholly within the realm of certainty, and avoid the realm of probability. There is, however, another barrier to reporting supplementary figures on reserves covering the "probable" and "possible." There is no uniformity in the systems which companies use for evaluating these categories, with the consequence that, even if the data were available, there would be serious problems in attempting to aggregate them.

As of January 1, 1965, the parent API Committee on Petroleum Reserves had the following membership:

Morris Muskat, *Chairman*, Gulf Oil Corporation, Coral Gables, Florida
D. V. Carter, *Vice Chairman*, Mobil Oil Company, Houston, Texas
S. A. Berthiaume, *Secretary*, Texaco, Inc., Houston, Texas
Joseph P. Buder, *Assistant Secretary*, American Petroleum Institute, New York City, New York
A. G. Copeland, Shell Oil Company, New York City, New York
T. A. Dawson, Indiana Geological Survey, Bloomington, Indiana
George H. Galloway, Pan American Petroleum Corporation, Tulsa, Oklahoma
Merrill W. Haas, Humble Oil & Refining Company, Houston, Texas
R. R. Lindsly, Phillips Petroleum Company, Bartlesville, Oklahoma
D. D. Little, Standard Oil Company of California, San Francisco, California
Kenneth E. Montague, Sun Oil Company, Beaumont, Texas
Raymond Rantala, The Pure Oil Company, Palatine, Illinois
B. L. Waggoner, Continental Oil Company, Houston, Texas

MISINTERPRETATION OF API PROVED RESERVES FIGURES

Because of the public importance of the future availability of oil, such data on reserves as there are receive a good deal of public attention. Analysts in companies, banks, government agencies and universities have something to say about the figures, and newspapers and magazines run articles for the general public.

There are two difficulties attending this wide attention. The first is that there are so few data. API estimates of proved reserves provide the only continuous series having an official and authoritative character. They therefore provide a starting point for all sorts of analytical exercises, which may, or may not, be justified by the character of the data, and which may or may not convey a correct impression, according to whether the writer and the reader understand the nature and limitations of the underlying figures.

Of all the calculations derived from API figures, the most widely known and commonly referred to is arrived at by dividing annual production into the aggregate proved reserves at the end of each year. To this ratio of proved reserves to production the names of "reserve life index" or "reserve ratio" are commonly assigned. The almost mysterious quality of this ratio is its remarkable stability over time. For the fifteen years, 1949 to 1964, proved reserves at year end varied only between 11.5 and 13.5 times the annual production.

At the higher end of the scale of ignorance about reserves figures this ratio is often interpreted to mean that, at present rates of production, all the known oil will be exhausted in eleven to thirteen years, and any further domestic supplies will be dependent upon future discoveries. The name of "reserve life index" is well-designed to foster this misapprehension. Anyone who has read the preceding pages of this paper will at once see how fallacious this interpretation is, since the economically recoverable oil content of existing fields is known to be much larger than "proved" reserves; and improved recovery methods will make the amount so much the larger.

Even though the life-index fallacy is easily disposed of, it will perhaps be worth while to quote some references of other writers to this ratio. Writing in 1950, Frederic H. Lahee said, "There has been a tendency—in fact, it has almost developed into a habit—among many of those people who bemoan the diminishing ultimate supply of oil reserves, to divide the estimated reserves as of the end of a given year by the quantity of oil produced in that year, and to state that the quotient represents the number of years left during which we may be able to benefit by our domestic supply. That this line of reasoning is fallacious may be seen if we apply it through a series of years."[9]

A. D. Zapp of the Geological Survey wrote, "The proved-reserve estimates have also been commonly misinterpreted as reflecting the true rate of increase in the quantity of petroleum producible from existing wells without change in production practice or as reflecting the full quantity that is economically producible from existing wells."[10]

A staff study for the Senate Committee on Interior and Insular Affairs says, "A common practice has been, and is, to express reserves in terms of the ratio between the reserves figure and current rate of production. For example, reported oil reserves are presently equivalent to about thirteen years of current production. . . . The public fearfully interprets this to mean that the United States will have run out of oil in thirteen years. . . ."[11]

A federal agency can, by using loose language in a report on the petroleum industry, help perpetuate the misconception. For example, in a report by the Attorney General the following language appears: "The concept of proved reserves . . . represents . . . *a conservative estimate* of the total amount of crude oil which will *ultimately be recoverable* over an indefinite period of years, assuming continuance of the present economic climate and techniques of recovery."[12] (Emphasis supplied.) The difficulty is that proved reserves cannot be concisely defined. They are not a "conservative estimate" of anything. They are an amount arrived at by following the specified definitions and procedures laid down by the API, and do not pretend to quantify the oil "ultimately recoverable" even from known fields.

Anyone interested in doing so can make various kinds of calculations from the API figures. It can be shown, for example, that gross additions

[9] "Our Oil and Gas Reserves: Their Meaning and Limitations," *Bulletin of the American Association of Petroleum Geologists*, Vol. 34, No. 6, June 1950.

[10] *Geological Survey Bulletin 1142-H* (Washington: Government Printing Office, 1962).

[11] *Report of the National Fuels and Energy Study Group on An Assessment of Available Information on Energy in the United States, to the Committee on Interior and Insular Affairs*, U. S. Senate. Committee Print, 87 Cong., 2 sess. (Washington: Government Printing Office, 1962.)

[12] *Fourth Report of the Attorney General* pursuant to Section 2 of the Joint Resolution of July 28, 1955, consenting to an Interstate Compact to Conserve Oil and Gas. Report as of September 1, 1959, p. 13. Further illustrating the confusion is a statement attributed to Secretary of the Interior Udall that ". . . the Nation's petroleum reserves constitute only a 12-year supply." *Washington Post*, May 13, 1964, reporting an address before the American Association for the United Nations.

to reserves of crude oil for the fifteen-year period 1945–63 added up to 43.1 billion barrels, or an average of 2.87 billion barrels per year. By five-year periods, the average annual additions were 3.24 billion for 1949–53, 2.75 for 1954–58, and 2.61 for 1959–63. This suggests a substantial stability in the rate of accrual of new reserves.

Or again, it can be calculated that "discoveries" were at an annual average rate of 586 million barrels for 1949–53, 452 for 1954–58, and 343 for 1959–63. This suggests a recent drop-off in the discovery rate.

Or again, it can be noted that *net* additions to proved reserves added up to 7.7 billion barrels for the fifteen-year period, broken down by five-year periods to 5.7 billion barrels for 1949–53, 1.6 for 1954–58, and 0.4 for 1959–63. These figures show that gross additions to proved reserves have been more than keeping pace with annual production, though with a declining margin and with negative changes in net crude oil reserves in three of the last five years. It is an interesting fact that, during the fifteen-year period, the largest net additions to reserves occurred in the five years when production was increasing most rapidly, from 1948 to 1953. There appears to be some sort of relation between production and additions to reserves; perhaps rising production stimulates the drilling which leads to large additions. Since 1956, production has hardly increased at all and this may be part of the reason for the declining rate of additions to reserves.

Calculations such as these have the surface appearance of providing interesting and important information concerning the trends of development in the oil industry. Within limits, when done by persons fully aware of the nature and limitations of the figures and when read by people with similar awareness, these exercises may be interesting and informative. Being, however, commonly used by people lacking this knowledge, the raw data and calculations from them commonly lead to unwarranted conclusions. As we shall show later, good trend analysis for the industry would require data reported in different ways from those used by the API in the estimation of proved reserves.

MISUSE OF RESERVES DATA IN COST ANALYSIS

Space does not permit a detailed listing of the innumerable misuses of reserves data both by uninformed persons writing about the industry and by industry spokesmen who should know better. It is important, however, to mention one glaring misuse because of its possible importance in policy making and in molding public and industry opinion. In an effort to compute the costs of finding and developing petroleum, the reserves added for a given year in the API estimates are frequently divided into the exploration, development, and production expenditures for that year. The result is called the "replacement cost per barrel" for oil. The preceding discussion should have made it clear that the reserves reported in any particular year are the result of expenditures made over a number of years in the past. The industry is mainly living on discoveries of several years ago. Similarly, expenditures made this year for such things as geological and geophysical work and exploratory drilling invariably do not bring forth reserves this year, but rather assist in the discovery of reserves in future years.

It is, of course, conceivable that an analyst working with company cost and reserves data could, by assigning costs and reserves to specific years and lagging the relationship where it is appropriate, come up with fairly representative cost per barrel of reserves figures for his company. Such studies are undoubtedly done, but rarely, if ever, published. To attempt to do such a study using national reserves and expenditure data can only create confusion.[13]

It is also possible to find statements in studies which relate drilling activity to additions to reserves. This type of study divides the number of barrels of API reserves added by the number of total or exploratory wells, or by the footage drilled in each category, to get the number of barrels found per well or per foot drilled. Such studies, if adequately explained and properly done with the necessary time adjustments in the data, can be informative. However, such data are almost invariably related in a superficial fashion to prove a point and rarely are the limitations of reserves data explained.

The end result of such abuses is that the same reserves data are often used to "prove" opposite views in discussions relating to fuel supplies, costs, incentives to invest, and the like. The uninitiated observer is likely to feel that the data are rigged or meaningless, or to be misled into conclusions which the data are insufficient to support. When the observers are policy makers, they can hardly avoid feeling puzzled and suspicious. The industry's reputation cannot but be damaged by careless, if not deliberately misleading, presentations.

The prerequisites of useful communication are a clear statement of definitions and procedures by those who provide data and a recognition of the limitations of the data and the uses to which they can be put by those who use them.

---

[13] See Wallace F. Lovejoy and Paul T. Homan, with Charles O. Galvin, *Cost Analysis in the Petroleum Industry* (Dallas: Southern Methodist University Press, 1964).

3

National Petroleum Council Reports on
Reserves and Productive Capacity

The API estimates of reserves, even if the figures are taken at face value, reveal nothing concerning the trend of discovery of "new" oil, in the sense of showing the later-revealed potentialities of fields discovered in particular years. For certain purposes, particularly those of public policy which we shall examine later, it would be of interest to attribute the later accruals of reserves back to the years of discovery of the fields in which they are located. From analysis of this sort—if the data were properly arranged— it would be possible to establish the trend (or absence of trend) in the additions to supply resulting from the fields originally discovered in each year. A limited contribution to analysis of this sort was originated by the Petroleum Administration for War (PAW) in 1945,[1] and updated in 1961

---

[1] The thinking of the PAW on the then existing reserves data cast some interesting light on the genesis of the entire dating-back process. Testifying before the Senate Special Committee to Investigate Petroleum Resources in 1945, one of the top PAW staff members noted:

> The American Petroleum Institute figures furnish an accurate and useful inventory of reserves actually proved by drilling, but they were not compiled to indicate the rate of discovery of new sources of supply.

> The Petroleum Administration necessarily was concerned principally with the rate at which new sources of supply were being discovered. Therefore, it began its own compilation of reserve estimates, and followed the method of crediting all reserves proved by development drilling back to the year in which the source of supply was discovered . . . (Philip H. Bohart, Director of Production Division, PAW, "Wartime Petroleum Production in the United States," *Petroleum in War and Peace*, papers presented by the PAW before the Senate Special Committee to Investigate Petroleum Resources, November 28–30, pp. 70–71).

and 1965 by the National Petroleum Council (NPC), a body made up of industry members which is advisory to the Secretary of the Interior. The NPC is co-chaired by an industry representative and a representative of the Interior Department.

The NPC has published a series of reports of its Committee on Proved Petroleum and Natural Gas Reserves and Availability on the productive capacity of the industry. In the 1961 and 1965 reports, at the request of the Department of the Interior, a special analysis of reserves was made, attributing later API extensions and revisions back to the year of discovery of the field or reservoir within which they were located.[2] These reports are of considerable importance for the clues they offer for trend analysis applicable to estimating future availability of petroleum. The discussion will draw on materials from both the 1961 and 1965 reports. In some respects the 1961 report contains a fuller statement of problems and techniques, and will be used in place of the 1965 report where it seems appropriate. We shall deal first with the analysis of reserves data and then with the estimates of productive capacity. In both parts of the reports, the analysis is applied separately to crude oil, natural gas, and natural gas liquids, but at this point we shall review only the crude oil portions.

NPC DATING OF RESERVES DATA

In its time series the API, it will be recalled, enters a very small figure for new discoveries in a single year, and then in that same year records the much larger addition to reserves based on "extensions and revisions in previously discovered fields." These latter additions in any year are derived from the changes in estimates of the reserves in all existing fields, no matter when they may have been discovered. They reveal nothing about the rate at which petroleum has been "discovered," as revealed by the evidence of later years. In its letter requesting that the analysis be made, the Department of the Interior stated the need for data which would throw light on "the trend of results obtained from exploration" and "the rate of discovery of new sources of supply."[3]

In carrying out its task, the NPC Committee set up regional subcommittees distinct from those of the API. There was, however, a considerable

---

[2] National Petroleum Council, *Proved Discoveries and Productive Capacity (1960)* (Washington: 1961); National Petroleum Council, *Proved Discoveries and Productive Capacity (1964)* (Washington: 1965). Hereinafter cited as *NPC Report, 1960*, and *NPC Report, 1964*.

[3] *Ibid., NPC Report, 1960*, Appendix A.

TABLE 4. SUMMARY OF ESTIMATED DISCOVERIES OF CRUDE OIL NOW ASSIGNED TO FIELDS
DISCOVERED IN YEARS SHOWN

(Thousands of barrels)

Year	District 1 [a]	District 2	District 3	District 4	District 5	Total U.S.
1919 [b]	2,006,106	6,232,032	3,922,944	787,455	5,642,781	18,591,298
1920	3,654	701,225	203,447	41,270	1,514,350	2,463,946
1921	3,327	435,103	1,633,717	85,050	879,000	3,036,197
1922	2,756	169,606	937,022	190,592	188,600	1,488,576
1923	1,858	588,551	329,109	43,073	279,000	1,241,591
1924	2,223	314,719	264,640	53,063	306,600	941,245
1925	2,368	76,494	735,509	61,141	155,700	1,031,212
1926	3,376	748,858	2,116,271	7,306	363,850	3,239,661
1927	4,000	578,889	508,018	206,148	169,496	1,466,551
1928	2,361	873,520	830,341	95,768	632,061	2,434,051
1929	4,500	525,675	2,245,285	14,650	66,435	2,856,545
1930	2,195	272,756	6,653,876	157,344	21,100	7,107,271
1931	730	270,357	1,619,602	6,300	537,404	2,434,393
1932	460	81,535	312,773	197,056	115,630	707,454
1933	740	304,780	375,306	602,254	86,040	1,369,120
1934	1,000	278,264	1,709,435	720	238,300	2,227,719
1935	1,500	674,669	2,692,781	25,830	44,610	3,439,390
1936	2,119	108,579	1,071,294	6,351	1,481,911	2,670,254
1937	8,900	1,006,185	3,477,135	54,450	284,016	4,830,686
1938	1,360	797,862	1,658,732	73,339	922,296	3,453,589
1939	990	325,138	975,135	0	111,614	1,412,877
1940	1,120	567,260	2,534,446	9,200	36,643	3,148,669
1941	1,230	592,153	1,101,859	0	487,561	2,182,803
1942	1,210	180,588	903,570	338,822	30,268	1,454,458
1943	13,769	304,675	907,142	167,053	49,340	1,441,979
1944	700	241,623	1,302,700	355,845	162,875	2,063,743
1945	2,365	129,741	1,567,300	16,100	206,941	1,922,447
1946	910	630,538	662,282	174,416	68,968	1,537,114
1947	2,262	258,438	599,328	37,466	265,835	1,163,329
1948	1,440	145,831	2,703,799	184,457	171,966	3,207,493
1949	900	198,702	2,054,073	191,924	279,039	2,724,638
1950	1,400	159,612	1,905,651	120,278	50,395	2,237,336
1951	860	236,095	785,928	187,802	69,758	1,280,443
1952	870	418,823	808,231	214,269	98,066	1,540,259
1953	740	163,446	1,262,014	243,833	56,692	1,726,725
1954	913	231,572	1,170,680	169,638	68,236	1,641,039
1955	980	193,258	848,527	81,162	88,969	1,212,896
1956	1,445	199,141	609,838	199,055	28,615	1,038,094
1957	1,340	268,669	623,612	169,913	196,951	1,260,485
1958	1,750	222,459	389,351	66,570	78,397	758,527
Total	2,092,727	20,707,421	57,012,703	5,636,943	16,536,309	101,986,103

[a] Composition of PAD districts is as follows:
   I—New England, New York, New Jersey, Pennsylvania, West Virginia, Virginia,
   North Carolina, South Carolina, Georgia, Florida, Delaware, Maryland.

overlap of expert personnel; the reserves concept which was utilized was substantially that of the API; and the cumulative additions to reserves which the NPC Committee manipulated differed but little from those of the API. The principal difference in totals was that the NPC, though operating from the vantage point of 1959, assigned only reserves in fields discovered during or before 1954 in the 1961 report. The 1965 report used 1963 data for fields discovered during or before 1958. The relation between the Committee's figures and those of the API may best be expressed in the Committee's own words in the 1961 report, as follows:

> The proved reserves of crude oil in the United States at the end of 1959 were estimated by the API to total 31.7 billion barrels. Cumulative production of crude oil as of the same date was approximately 62.6 billion barrels, indicating total discoveries of crude oil from 1859 through 1959 of 94.3 billion barrels as estimated by the API at the end of 1959. The historical tabulation presented in Table 1 of this report shows 91.4 billion barrels of the total oil discovered through 1959 to be assigned to fields discovered through 1954. The remaining 2.9 billion barrels represents the 1959 estimate of recoveries from fields discovered during 1955-59, together with the net total of any differences in the estimated recoveries between the API Committee and the NPC Committee.

It is therefore substantially correct to say that the NPC reports are basically a temporal reallocation of cumulative proved reserves as estimated by the API. Included in the NPC figures are, of course, all past production. Since, as the Committee points out, the fields surveyed may be expected to yield large additions to proved reserves in the future, the data presented fall short of giving a full picture of the trend of discoveries. The methodology of the analysis followed closely a similar report which had been made for the PAW in 1945.

The summary figures for crude oil from the 1965 report are presented in Table 4. The effect of reassigning reserves to the year of discovery of

---

II—Ohio, Indiana, Illinois, Michigan, Wisconsin, Minnesota, Iowa, North Dakota, South Dakota, Nebraska, Kansas, Oklahoma, Missouri, Tennessee, Kentucky.
III—New Mexico, Texas, Louisiana, Mississippi, Alabama, Arkansas.
IV—Montana, Idaho, Wyoming, Utah, Colorado.
V—Washington, Oregon, California, Nevada, Alaska, Arizona, Hawaii.
b 1919 includes all years previous.

SOURCE: National Petroleum Council, *Proved Discoveries and Productive Capacity (1964)* (Washington: 1965).

fields may be seen by comparing the figures with those of Table 2 in Chapter 2. From 1946 to 1963 the API credited 43.9 billion barrels to "revisions and extensions," and 8.0 billion to "discoveries." As compared to this total of 51.9 billion barrels, the NPC credited fields discovered from 1946–58 with only 21.3 billion barrels of reserves "proved" during 1946–63. In other words, a major fraction of reserves proved, 1946–63, were in fields discovered before 1946.[4] In parallel fashion, a large fraction of reserves in fields discovered after 1946 will be proved in years after 1963.

In summarizing its 1960 findings, the Committee says: ". . . the estimated recovery from the average field usually increases several fold over the estimate made at the end of the discovery year and continues to increase from various causes for a great many years. . . . Studies of the long-term future supply of oil in the United States must give cognizance not only to the future possible discoveries of new fields but also to the future additions to reserves from existing fields."[5] ". . . a much larger quantity of additional oil may be expected in future years to be recoverable from existing fields than is reflected by the present estimates of proved discoveries. This additional oil will come from extensions of existing fields by future drilling, by upward revision of recovery estimates as more knowledge of the reservoirs is obtained, and by increasing recoveries through application of improved recovery techniques."[6]

The same point is illustrated in the 1965 Report by Table 5, which shows a comparison of API discoveries and 1963 estimates of discoveries for fields in different time periods. The ratios of present to initial estimates are much greater as we go back in time.

TABLE 5. COMPARISON OF PRESENT ESTIMATES OF DISCOVERIES WITH INITIAL API ESTI-
MATES FOR NEW FIELDS AND NEW POOLS

(Millions of barrels)

Fields discovered during period	Initial API est. of discoveries	Present est. of discoveries	Ratio: present to initial est.
1939 through 1943	1,600	9,686	6.1
1944 through 1948	2,017	9,888	4.9
1949 through 1953	2,933	9,525	3.2
1954 through 1958	2,261	5,859	2.6

SOURCE: National Petroleum Council, *Proved Discoveries and Productive Capacity (1964)* (Washington: 1965), p. 7.

[4] This is a very rough comparison, sufficient to make its point, but not covering the precise steps necessary to reconcile API with NPC figures. The general terms of this reconciliation will be found on p. 12 of the 1961 NPC report.

[5] *NPC Report, 1960*, p. 3.

[6] *Ibid.*, p. 13.

Though recognizing the large amount of reserves which will in the future be added to the "proved" estimates for existing fields, the Committee limits itself to the consideration of proved reserves. It does not suggest any statistical procedure by which to estimate the unproved content of existing fields. However, as we shall see, if the proved additions were continuously recorded year by year back to the year of discovery of the field, it would not be difficult to derive a factor by which to make such projections, within certain limiting economic, technological, and geological assumptions.[7]

The Committee is careful to point out that its own procedures do not provide a sufficient basis for true trend analysis.

> Because of the growth in estimated recovery as more is known about a field and technology improves, a proper analysis of discovery trends would require comparison of annual discoveries estimated by comparable methods at equal intervals after discovery. . . . the historical tabulations presented in this report are not so constructed but present, instead, a single set of estimates, all based on data available at the end of 1959. The fields discovered in 1954 have been credited with only five years' revisions, whereas those discovered in earlier years have been credited with revisions for progressively longer intervals. . . . The resultant distortion is far too great to permit sound deductions with respect to discovery trends from these tabulations alone.[8]

This kind of explicit statement of the limitations of the figures, and of proper interpretation of them, is something that might well be copied in the API annual reports on proved reserves, since the figures from those reports are subjected to endless misinterpretation and misuse.

TABLE 6. COMPARISON OF ESTIMATES OF DISCOVERIES OF CRUDE OIL TABULATED ON SIMILAR BASES

(Millions of barrels)

Fields discovered during period	PAW estimate made in 1945	NPC estimate made in 1960	NPC estimate made in 1964
Total through 1919	14,640	17,367	18,591
1920 through 1944	37,482	56,244	60,219
1945 through 1954	—	17,815	18,949

SOURCE: National Petroleum Council, *Proved Discoveries and Productive Capacity (1964)* (Washington: 1965), p. 7.

[7] It might be noted in passing that the NPC procedures and data require reports on reserves by fields over an extended period. It seems reasonable to assume the NPC Committee did not make extensive field-by-field calculations but rather relied on API subcommittee working papers.

[8] *NPC Report, 1960*, pp. 3-4.

The 1965 report provides some striking illustrations of the inexorable forces at work over long time periods. One, given below in Table 6, shows how estimates of "discoveries" in the *same* fields, but made at different points in time, can differ greatly.

The 1964 estimate for fields discovered before 1920 was 1.2 billion barrels greater than the estimate for the same fields made in 1960, and almost 4 billion barrels greater than the 1945 estimate. In the Committee's words in 1961:

> The Committee is well aware, of course, that the increases in esti-
> mates of recoverable oil from fields previously discovered do not
> take place automatically with the passage of time, but, instead, are
> the direct result of further drilling, the development of additional
> information, and the application of improved recovery procedures.
> The rates at which these activities take place may be accelerated or
> slowed down, depending upon the prevailing opportunities, economic
> incentives, and other influences, including governmental policies. The
> comparisons provided in this report might be more specifically inter-
> preted, therefore, as follows:
>
> 1. The estimated proved recovery from newly discovered fields,
> when estimated at the end of the discovery year by the concepts and
> methods commonly employed in the petroleum industry, usually
> represents only a very small fraction of the oil that will ultimately
> prove to be recoverable.
>
> 2. The average newly discovered field becomes a new source of a
> continuing stream of additions to reserves, the growth in estimated
> recovery being reflected in extensions and upward revisions.
>
> 3. The growth in estimated recovery from the average field does
> not cease after a few years, but continues for a very long period.
> Comparisons between the estimates in the present report and other
> earlier studies show that growth is still taking place in estimated
> recoveries of fields over 40 years old.
>
> 4. The aggregate of all fields heretofore discovered, which is com-
> prised of a distribution of fields in all states of development, from
> newly discovered ones to those a century old, represents an enormous
> base for furture additions to reserves through further extensions and
> revisions.
>
> 5. The ultimate magnitude of future extensions and revisions of
> estimated recoveries from presently existing fields has not been deter-
> mined. In principle, it must be limited to complete recovery of the oil
> in place. However, the available data do not indicate that this
> theoretical limit is, as yet, a restrictive influence on growth in recovery
> estimates.

6. It is evident from the foregoing considerations that studies of the long term future supply in the United States must give cognizance not only to the possible future discoveries of new fields but also to the future additions to reserves from existing fields.

7. The rate of discovery of new fields and the rate of increase in estimated additions to reserves of existing fields are not simple functions of time but are related to the intensity of the effort expended in the search for oil, in both exploratory and development drilling, and in technologic development and its application through improved recovery methods. The intensity of these activities depends, in turn, on the need as reflected in growth in demand, and on financial incentives. There is a crucial interdependence among these factors. It is beyond the scope of this report, however, to comment on this complex but decisive interrelationship.[9]

For persons wishing to inform themselves on the subject of reserves, and heretofore limited to API sources, the explanatory text of this report is invaluable. It both extends the usefulness of API figures and helps guard against their misuse. It suffers, however, from one serious limitation. It gives no direct clue to reasonable expectations or how to arrive at them about the reserves ultimately recoverable on economic terms from existing known fields. It sticks to the "facts" of proved reserves. It is encouraging, however, to note the Committee's heavy emphasis on economic incentives in helping to determine the manner in which reserves come into being and the rate at which they are found and developed.

It is not a criticism of the NPC reports that they do not go into the problems of quantifying these more generous expectations. The Committee operated within specific terms of reference and well fulfilled its function within those terms. However, to arrive at reasonable expectations concerning future availability of oil, it is necessary to have data and to devise analytical procedure which lie outside the scope of the NPC reports.

RESERVES ESTIMATION FROM EXPLORATORY DRILLING DATA

Since the interesting feature of the NPC procedure was that of backdating proved reserves to the year of original discovery of the fields within which they were located, another study using a backdating technique may

---

[9] *NPC Report, 1960*, pp. 17-20. A similar but much abbreviated statement is found in the *NPC Report, 1964*, p. 8.

be usefully interpolated here. Frederic H. Lahee, and others following him, have used exploratory drilling statistics for seventeen major producing states reported by the American Association of Petroleum Geologists (AAPG) and have determined a size distribution of oil and gas fields found in the past from the results of exploratory drilling effort. [10]

Each year the AAPG reports (1) the number of exploratory wells drilled by states (or part states), (2) whether the wells find oil, gas or condensate in commercial quantities or in noncommercial quantities, and (3) whether the wells are dry. Since success rates of exploratory drilling tell only part of the story and do not explain the quantities found in each year's successful wells, Mr. Lahee attempted to bridge the difficult gap between exploratory effort and tangible results, that is, reserves. Of particular significance for our purposes is the reporting of fields by size after six years of development. Oil and gas fields are ranked according to size in the following manner:

Field Class	Field Size (ultimate recoverable reserves)
A	300 million MCF or more of gas; 50 million Bbls. of liquids or more; or a combined equivalent;
B	150-300 million MCF of gas; 25-50 million Bbls. of liquids; or a combined equivalent;
C	60-150 million MCF of gas; 10-25 million Bbls. of liquids; or a combined equivalent;
D	6-60 million MCF of gas; 1-10 million Bbls. of liquids; or a combined equivalent;
E	Less than 6 million MCF of gas; less than 1 million Bbls. of liquids; or a combined equivalent; and
F	Fields abandoned as non-profitable.

Using data for "new field wildcats" only for a seventeen-state area (accounting for 95 per cent or more of exploratory drilling, oil reserves, gas reserves, oil production, and gas production), Mr. Lahee examined, in 1957, the results of wildcat drilling for the years 1943 through 1952. Carsey and Roberts have updated the initial study and have added 1953–57

[10] Frederic H. Lahee, "Statistics on Natural-Gas Discoveries," *Bulletin of the American Association of Petroleum Geologists*, Vol. 42, No. 9 (September 1958), p. 2037; and *ibid.*, "Trends in Exploration," *Proceedings of the American Petroleum Institute*, Division of Production, Vol. 37 [IV], (1957), p. 38; J. Ben Carsey and M. S. Roberts, "Exploratory Drilling in 1962," *Bulletin of the American Association of Petroleum Geologists*, Vol. 47, No. 6 (June 1963), p. 889; L. R. Newfarmer and E. L. Dillion, "Exploratory Drilling in 1963," Vol. 48, No. 6 (June 1964), p. 749.

to Lahee's earlier years. They find the following array of fields ranked by size and combining oil and gas reserves on an equivalent Btu basis:

Number of Fields Discovered, Classified by
Estimated Total Ultimate Recoverable Reserves
after Six Years of Development History [a]

	A	B	C	D	E	F	Total
1943	7	9	17	69	200	25	327
1944	6	3	27	67	141	36	280
1945	6	3	23	77	139	52	300
1946	3	3	17	72	149	44	288
1947	8	12	15	82	164	67	348
1948	5	5	16	92	276	52	446
1949	17	11	17	99	215	90	449
1950	7	14	23	97	327	75	543
1951	11	5	23	112	402	77	630
1952	12	9	18	114	426	105	684
1953	4	9	23	128	372	152	688
1954	6	4	24	133	412	218	797
1955	4	8	24	128	436	221	821
1956	2	6	7	144	425	150	734
1957	10	6	22	133	444	216	831

[a] Development history for gas wells not necessarily six years.

It is to be noted that this distribution would change as revisions and extensions prove up additional reserves. Fields in the D category in 1952 might move to the C or B category in a later year if more than six years of development history were used. Or conversely, on additional evidence, fields may move into a lower category.

The Lahee, *et al.*, technique of determining the frequency distribution of oil and gas fields by size of reserves measured after several years of production and development history might very well be adapted to solve the problem of estimating near future reserves from current exploratory drilling statistics. Admittedly, the precise amount of oil and gas that would be discovered would not be predictable, but more work along these lines, using the AAPG data, would contribute to trend analysis. Further refinements in the data to determine more precisely the average size field in each category, and development of some moving average techniques to smooth the series is called for. The exploratory drilling statistics are as yet a relatively untouched source of information relating to reserves data. Since the Lahee type of analysis implies information on reserves, and on changes in reserves, by fields, such information could presumably be adapted to analysis for determining the rate of growth of reserves, adjusted to years of discovery.

NPC ESTIMATES OF PRODUCTIVE CAPACITY

Portions of the 1961 and 1965 NPC reports are devoted to estimating the productive capacity of existing crude oil wells (and also natural gas and natural gas liquids capacity). This part of both reports is a continuation of a series of intermittent reports since 1951, prepared by committees of the NPC. The latest estimates are of productive capacity as of January 1, 1964, broken down by the five Petroleum Administration for Defense (PAD) districts.[11] Productive capacity figures by states, and even by subdivisions of major producing states are compiled by the NPC Committee, but these data have only limited circulation and are not published. They are, however, provided to the Independent Petroleum Association of America (IPAA) to assist in its annual estimates referred to below.

The crude oil productive capacity figures from the various NPC reports on a national scale for the period 1951–64 are as follows:

January 1, 1951	6,727 thousand barrels per day
January 1, 1953	7,465 thousand barrels per day
July 1, 1954	8,331 thousand barrels per day
January 1, 1957	9,367 thousand barrels per day
January 1, 1960	10,585 thousand barrels per day
January 1, 1964	11,590 thousand barrels per day

The addition of natural gas liquids brings the 1964 total to 14,393 thousand barrels per day.

The Committee is careful to point out that the productive capacity concept is vastly different from a concept of "current availability" or "deliverability." The definition of productive capacity used by the NPC Committee is as follows:

Productive capacity as used in this report represents an estimate of the aggregate capability of the various fields and reservoirs in the United States within the limitations of generally accepted production practices, but viewed primarily from the standpoint of the physical

[11] For a description of the PAD districts, see notes to Table 4.

capabilities of the fields and reservoirs and not, in general, from the standpoint of the limitations that might be imposed by lack of markets, lack of transportation, inadequacy of mechanical equipment, lack of facilities for proper handling of gas or water produced with crude oil, or any other correctible impediments to production.[12]

The Committee takes pains to point out that its estimates of productive capacity are based on (1) actual production in fields which are currently producing at capacity, and (2) *estimates* of what restricted (by state regulation or market factors) fields could produce efficiently. The tremendous number of fields and reservoirs makes a complete and detailed study of every reservoir infeasible. For many reservoirs, capacity estimates are based on comparisons of fields with similar physical and operating characteristics. "The figures reported represent *estimates* based on informed and experienced judgement." (Emphasis in the original.)

The Committee is also careful to point out that its estimates of crude oil producing capacity *exclude* condensate. Condensate capacity is included with estimates of natural gas liquids. This being the case it is not entirely accurate to compare NPC crude oil producing capacity with Bureau of Mines crude oil production figures, since the latter includes some "lease condensate." The 1965 U.S. Interior Department study of the petroleum industry reported that lease condensate was about $4\frac{1}{2}$ per cent of total crude and condensate production as reported by the Bureau of Mines in the early 1960's.[13] The implication is that anyone computing the percentage of capacity currently utilized by using capacity and production figures should deflate the results by some factor. In other words we are using less of our capacity than would appear from a comparison of Bureau of Mines production and NPC capacity data.

The 1961 Report contains estimates of what capacity would be available in the future, given several assumptions about drilling rates. The 1965 Report contains no such estimates, although the Interior Department has subsequently requested this information.[14] The estimates of this sort in the 1961 Report provide some interesting insights into the future posture of the industry.

In 1961 the Committee estimated that, assuming no new productive drilling or reservoir stimulation, if these wells were produced at a capacity rate, at the end of one year the capacity would be reduced from the

[12] *NPC Report, 1964*, p. 13.
[13] *An Appraisal of the Petroleum Industry of the United States* (Washington: 1965), Table 28.
[14] NPC News Release, April 23, 1965.

January 1, 1960 figure of 10,585 thousand to 9,694 thousand barrels per day, or an average annual decline rate of 8.4 per cent. At the end of the second year the daily capacity would be 8,905 thousand barrels, or an average annual decline rate of 8.1 per cent.

It is further estimated that, if the wells were produced at capacity, it would be necessary to drill from 20,000 to 25,000 successful oil wells annually (not counting dry holes) to maintain capacity at the level of 10,585 thousand barrels per day. Since recent experience is roughly one successful completed oil well to every two drilled, this implies that if production were at capacity rates, some 40,000 to 50,000 wells *in toto* must be drilled annually to maintain the 10,585 thousand barrels of daily capacity.[15] A large proportion of successful wells is, of course, drilled as development wells on already proved territory, but the total would have to include a substantial number of successful wildcat wells to provide new areas for development.

The drilling estimate is based on a number of specific assumptions.

It implicitly assumes the continuation of exploration at the level required to support the necessary wildcat drilling. It also assumes that during the period under consideration there will be no material change from the recent past with respect to geographic distribution of wells, quality of reserves developed, proportion of productive oil wells to total wells, or spacing of wells, and that future oil wells will, on the average, develop new productive capacity at about the same rate as the wells heretofore drilled.[16]

Changes in any of these assumptions—and some such changes are likely to be called for—could change the drilling requirements substantially.

Since the present rate of actual production is only about two-thirds the capacity rate, existing capacity could be maintained by a lower rate of drilling. Or, to put the matter another way, drilling at the indicated rate would result in the maintenance of excess capacity until rising demand absorbed the excess.

A separate estimate of productive capacity has been made annually since 1954 by the Productive Capacity Committee of the IPAA, distributed among the five PAD producing districts and based on data for each state. As of January 1 for recent years, the IPAA capacity figures run as follows, in thousands of barrels daily:[17]

[15] Included in the other 20,000 to 25,000 wells are dry holes, gas wells, and service wells.

[16] *NPC Report, 1960*, p. 29.

[17] Report of the Productive Capacity Committee, Independent Petroleum Association of America, Midyear Meeting, Denver, May 2-4, 1965 (mimeographed).

	1961	1962	1963	1964	1965
Crude oil and lease condensate	9,892	10,081	10,169	10,286	10,534
Processed natural gas liquids	1,041	1,049	1,090	1,177	1,222
Total petroleum liquids	10,933	11,130	11,259	11,463	11,756

The IPAA figures regularly run somewhat lower than those of the NPC, the difference being due in part to a difference of definition which is explained by a participant in the IPAA procedures as follows:

With regard to reconciling the IPAA and NPC figures, I don't think, overall, there is too much difference when you consider the difference in definition. The IPAA defines productive capacity as "the average rate of production from existing wells that could be maintained for a period of from 6 to 12 months without further development and with no significant loss in ultimate recovery." As I understand it, the NPC figure is more of an instantaneous capacity, which might be reached on a specific day but could not be maintained without further drilling. If you apply a normal decline rate to the NPC figure for January 1960 and assume no drilling, the figure at the end of 1960 probably would approximate the IPAA figure for January 1960 which, according to our definition, could be maintained for 6 to 12 months with no drilling.

Another difference between NPC and IPAA data results from varying definitions of crude oil. It was noted above that the NPC reports producing capacity for crude oil only and excludes lease condensate. The IPAA, on the other hand, includes lease condensate in its capacity figures for crude. Thus if we exclude lease condensate from the IPAA figures, the result would be an even lower figure for the IPAA compared to that of the NPC for any year.

It is somewhat puzzling that the NPC and IPAA figures diverge so greatly. If we take the 1964 NPC crude capacity figure of 11,590,000 barrels daily and reduce it by about 6 per cent, which is roughly an 8 per cent decline rate for nine months, we get a figure of about 10,894,000 barrels daily. This should give something less than the IPAA figure, using the IPAA definition of capacity producible for from six to twelve months, since the IPAA includes lease condensate. The reverse is true, however. The 1964 IPAA figure is 10,286,000 barrels daily, *including* some lease condensate which is not in the NPC definition. If we use the $4^1/_2$ per cent figure from the Interior Department study noted above as the amount of lease condensate capacity in the combined lease condensate and crude oil capacity, we may then deflate the IPAA figure to get something comparable to the NPC figure. The result is a 1964 IPAA capacity figure for crude minus lease condensate of 9,823,000 barrels daily (10,826,000 − 4.5 × 10,285,000). This is more than a million barrels of daily crude oil producing capacity less than the 10,894,000 barrels daily which we estimated for the NPC using a decline rate to achieve comparability between definitions. We are unable to reconcile the two series on productive capacity.

Since the IPAA does no estimating of reserves, and since productive capacity figures must rely heavily on reserves data, it appears that the NPC capacity figures constitute the primary source of such information. The IPAA Committee can provide considerable continuity in the series by its annual reports for it has access to the NPC capacity breakdown by states and also has access to the published API data on reserves, as well as data on production, and the number of wells drilled, producing wells, and abandoned wells by states.

There still remains a major gap in both the NPC and IPAA series, for no description is given of the methods used to determine the productive capacity figures. "Productive capacity" is a highly ambiguous concept. It can mean wells wide open, or reservoirs produced at an efficient rate (MER), or at some other assumed rate. It also can take account of the practices of state regulatory agencies which adjust the rate of production between various producing interests. The basis of estimation needs to be carefully defined.

Figures on productive capacity are, of themselves, not obviously relevant to the subject of reserves which we are investigating. But they become so in the hands of investigators who use producing capacity as one of the instruments for estimating recoverable reserves. This use will be illustrated in a later section when we review the Geological Survey study by A. D. Zapp.

4

Extending the Boundaries of Reserves Estimation: The IOCC Estimates

From what we have seen about the API estimates of proved reserves, it is evident that they represent only a fraction of the oil which well-informed experts expect to be recoverable from known fields, to say nothing of undiscovered fields. They represent a specifically defined measure of the working inventories of oil immediately available to producing companies, but do not reflect the expectations of additional amounts which will in the future be proved in presently known fields, within ranges of relative certainty or uncertainty.

Naturally, for a variety of purposes, there is a desire to extend the knowledge of oil resources to include reasonable expectations of recovery from existing known fields. A limited contribution to the extension of knowledge in this direction is the series of reports issued at two-year intervals between 1954 and 1962 by the Secondary Recovery and Pressure Maintenance Advisory Committee of the Interstate Oil Compact Commission (IOCC), under the guiding direction of Paul D. Torrey, chairman of the Subcommittee on Oil Resources. The primary purpose of these reports has been to establish reasonable expectations concerning the amount of crude oil which may be recovered from known fields by applying fluid injection methods under present economic conditions. But beyond this, they present data which open up a wider field of speculation concerning future recovery.

The master table from the fifth biennial report,[1] as of January 1, 1962,

[1] Paul D. Torrey, "Evaluation of United States Oil Resources as of January 1, 1962," *Oil and Gas Compact Bulletin*, Vol. XXI, No. 1, June 1962.

is reproduced in Table 7. The basic data from the table may be summarized as follows:

1. Original crude oil content of known reservoirs — 346.2 billion barrels
2. Total crude oil production to 1/1/62 — 67.8 billion barrels
3. 1961 crude oil production — 2.5 billion barrels
4. API proved reserves as of 12/31/61 — 31.8 billion barrels
5. IOCC primary reserves as of 1/1/62 — 31.4 billion barrels
6. Estimated additional recovery by conventional fluid injection methods under economic conditions as of 1/1/62 — 16.3 billion barrels
7. Estimated additional reserves physically recoverable by known methods — 40.2 billion barrels

PRIMARY RESERVES

It will be noted that item 5 presents an estimate of "primary" reserves very nearly the same as the API estimate of "proved" reserves in item 4. The sources of information are, however, different, and the two figures are, in principle, differently defined. The API figure includes all the reserves based on secondary recovery methods which they classify as "proved," while the IOCC "primary" figures specifically exclude such reserves, carrying them instead into item 6. Concerning sources, the report states that "Most of the member states of the Compact now have effective organizations composed of representatives from industry and from the regulatory authorities who have designated responsibility for the development of information on which oil resources studies can be based."

In further comment upon the methods of estimation of primary reserves, Mr. Torrey has stated:[2]

> The method of estimating primary reserves by the API Committee and by the Compact Committee to some extent is similar. However it will be understood that for a majority of the important oil producing states the members of the Compact Committee are either connected with a state agency or authority or have access to official information by reason of their membership in the Compact Committee. Thus, the opinion can be expressed that the information available

[2] In private correspondence.

TABLE 7. EVALUATION OF U. S. OIL RESOURCES AS OF JANUARY 1, 1962

(Millions of barrels of 42 U. S. gallons)

State	(1)	(2)	(3)	(4)	(5)	(6)	(7)
New York	855	208	2	28	0	20	210
Pennsylvania	3,831	1,236	6	102	7	77	904
West Virginia	2,610	467	3	51	38	13	585
Kentucky	1,473	437	18	116	138	55	65
Ohio	4,692	672	6	76	75	5	380
Indiana	3,080	328	11	62	45	50	300
Illinois	7,471	2,307	77	503	302	765	2,050
Michigan	1,822	461	19	79	97	129	444
North Dakota	3,000	126	24	413	448	147	574
South Dakota	2	1	a	b	1	0	0
Nebraska	1,603	166	24	100	59	119	150
Kansas	16,150	3,421	112	878	820	110	1,500
Oklahoma	49,715	8,396	188	1,787	1,420	1,331	4,560
Arkansas	5,666	1,135	28	281	235	158	316
Alabama	358	45	7	36	106	100	111
Mississippi	3,425	741	53	401	450	190	1,400
Louisiana c	27,346	5,571	365	4,931	5,187	4,416	4,525
Texas c	117,700	24,701	898	14,850	15,518	4,954	11,081
New Mexico	11,048	1,614	110	1,090	990	936	3,743
Colorado	3,266	645	47	420	164	613	1,301
Wyoming	11,482	2,063	143	1,380	969	626	1,372
Montana	2,853	394	31	251	216	118	116
Utah	1,797	152	33	218	170	276	347
California c	64,012	12,353	299	3,615	3,650	1,010	4,120
Alaska	776	7	6	b	250	100	0
Other States d	162	10	1	90	44	14	29
Total	346,195	67,657	2,511	31,758	31,399	16,332	40,183
Oil Resources as of 1/1/60 (4th Report)	328,407	62,636	2,483 e	31,719	30,970	14,822	44,013 f
Difference 4th to 5th Report	+17,788	+5,021	+28	+39	+429	+1,510	−3,830
Difference 3rd to 4th Report	+18,500	+4,818	−76 g	+1,419	+390	+1,727	+9,530 f

Identification of Column Numbers

(1) Estimation of orginal oil content of reservoirs.
(2) Total oil production to 1/1/62.
(3) 1961 oil production.
(4) API proved reserves, from Volume 16, American Petroleum Institute and American Gas Association, "Proved Reserves of Crude Oil, Natural Gas Liquids, and Natural Gas, December 31, 1961."
(5) IOCC primary reserves as of 1/1/62.
(6) Estimated additional recovery by conventional fluid injection methods under economic conditions as of 1/1/62.
(7) Estimated additional reserves physically recoverable by known methods, including solvent extraction and thermal.

Footnotes

a 0.2 million barrels.
b Included with other states.
c Includes offshore reserves.
d Includes Arizona, Florida, Missouri, Nevada, Tennessee, Virginia, and Washington.
e 1959 oil production.
f Oil recovery by means of solvents and heat was first considered in the fourth biennial report and was not considered in the third report.
g Change from 1957 to 1959 Production.

SOURCE: Paul D. Torrey, "Evaluation of United States Oil Resources as of January 1, 1962," *The Oil and Gas Compact Bulletin*, Vol. XXI, No. 1 (June 1962), p. 16.

to the Compact Committee on proved primary reserves is just as complete and just as reliable as is the information available to the API Committee. The primary reserve estimated for each state is based on a projection or on a calculation of the anticipated future production from every oil field. The annual reports of the Illinois Geological Survey and of the Louisiana Department of Conservation are cited as typical examples of the way in which the data are assembled. I believe that the estimations of primary reserves made by state authorities are just as reliable as the estimations of the API Committee. . . . There is bound to be some overlap in the sources of information available to the two committees, but, as far as the Compact is concerned, we have developed information on primary reserves in our own way. The fact that our figures for primary reserves correspond fairly closely with the API figure for proved reserves is just a coincidence.

RESERVES BASED ON FLUID INJECTION

The reserves attributed to fluid injection and economically recoverable under present conditions amount to 16.3 billion barrels. Added to the IOCC estimate of primary reserves, they give a total of 47.7 billion barrels, or 50 per cent more than API proved reserves. With respect to this supplementary estimate, Mr. Torrey states:[3]

It is my belief that additional reserves from conventional fluid injection operations can be estimated as accurately as primary reserves can be estimated. In each case the quality of the estimation is dependent on the quality of the data on which the estimation is based. The API Committee has underestimated reserves consistently and that is why the Compact Committee has been endeavoring to give governmental authorities more realistic information on which projections of future availability of oil can be based.

It is perhaps not accurate to say that the API "underestimates" reserves; rather it estimates a strictly defined body of reserves. But the IOCC estimates do reflect the urge in various quarters to expand the range of estimation with a view to greater knowledge of "future availability."

There is one area of vagueness in the IOCC Report which is not of great significance today but may be quite important in the future. In Table 7,

[3] In private correspondence.

item 5 is referred to as primary reserves and item 6 as recovery from fluid injection. It is not made clear how reserves are classified for fields in which pressure maintenance operations are begun early in the life of these fields. Such operations usually involve fluid injection; yet these fields clearly have primary reserves which would be producible without such injection. The dividing line between primary and secondary reserves in such situations becomes exceedingly hazy. It is not clear whether item 5 contains any reserves attributable to pressure maintenance, or whether all such reserves are carried to item 6. It seems likely that estimates of production decline curves without pressure maintenance are made to form a basis for dividing primary and secondary reserves in these ambiguous situations.

Separately listed is a special class of potential reserves described as "estimated additional reserves physically recoverable by known methods, including solvent extraction and thermal." The potential additional reserves attributed to this source are 40.2 billion barrels. In addition to known methods not yet commercially utilized, this source is said to include "a lot of oil that is physically recoverable by conventional methods of gas and water injection but which cannot be produced profitably under existing economic conditions." Mr. Torrey has amplified the statement as follows:[4]

Included in the 40.2 billion barrels is approximately 15 billion barrels of oil that is physically recoverable by known methods of fluid injection but which cannot be produced profitably under present economic conditions. The remainder, or some 25.2 billion barrels, is oil that is susceptible to recovery by solvent action, by thermal methods, etc. It should be understood that we consider these figures to be speculative and we had reason to reduce them radically in our last report by reason of the failure of certain miscible projects.

OIL ORIGINALLY IN PLACE

A unique feature of the IOCC report is the estimate of oil originally in place in known fields, the only estimate of this sort from any source. The estimated amount is 346.2 billion barrels. Since cumulative production to date had been 67.7 billion barrels, the estimated present oil content is 278.5 billion barrels. These figures can serve as the basis of various sorts of calculations. For example, recovery to date represents less than 20 per

---

[4] In private correspondence.

cent of estimated original oil in place. If to recovery to date is added the estimated "primary" and "secondary" reserves of 47.7 billion barrels, the resulting recovery would be about 33 per cent. Each percentage point added to the recovery rate by improved recovery methods or economic factors would add about 3.5 billion barrels to estimated reserves at current reserves levels.

Reservations are expressed in expert circles about the amount of credence to be given the estimate of original oil in place. Since, however, it is the only estimate there is, it is widely used as a basis for calculating future possible additions to oil supply which may become available through improvements in the technology of recovery or under suitable economic conditions. Concerning the reliability of the estimate, Mr. Torrey has stated:[5]

> In almost every state the reliability of the figures for all of its fields is not a constant. For fields of recent discovery, where fairly complete reservoir data are available the figure for original oil in place represents a volumetric calculation. For the older fields such complete data many times are not available, except where cores of the reservoir may have been taken in connection with secondary recovery operations. Therefore, in such cases we have to resort to estimations based on production performance. If, for instance, we know that a field has produced one million barrels of oil by dissolved gas drive it would be a reasonable assumption based on average dissolved gas drive performance that its reservoir originally contained six and two thirds million barrels of oil. We know that this is a backhanded method of calculation and that the judgment and the knowledge of the person making it is most important. Although we realize that our figures for individual fields might be questioned, when they are all combined into a single figure for a state such as Louisiana I believe that we come up with a reasonable figure.

The U.S. Department of the Interior has reported a new estimate by the U.S. Geological Survey of oil in place in known fields; the estimate is over 400 billion barrels in 1964, a substantial increase over the 1962 IOCC figure.[6] Details of the new study had not been published at the time of writing.[7]

---

[5] In private correspondence.

[6] U.S. Dept. of the Interior, *An Appraisal of the Petroleum Industry of the United States* (Washington: 1965), p. 15.

[7] Since the manuscript for this study was completed, an interesting report has been published which is relevant to a discussion of secondary reserves, or the recovery of oil

SOURCES OF IOCC DATA

The state sources for the IOCC estimates are for the most part officials of state agencies, but include a few independent or company-connected experts, and two connected with the U.S. Department of the Interior. The list for the fifth biennial report is as follows:

New York	Arthur M. Van Tyne, Geologist, New York State Geological Survey
Pennsylvania	William S. Lytle, Assistant State Geologist, Pennsylvania Geological Survey
West Virginia	A. J. W. Headlee, Geochemist and Chairman, West Virginia Secondary Recovery Committee
Kentucky	Edmund Nosow, Geologist, Kentucky Geological Survey
Ohio	Jack Cashell, President, The Preston Oil Company
Indiana	H. R. Brown, Director, Division of Oil and Gas, Indiana Department of Conservation
Illinois	Carl W. Sherman, Head, Petroleum Engineering Section, Illinois Geological Survey Division
Michigan	Gordon H. Hautau, Petroleum Geologist-Engineer, Geological Survey Division, Michigan Department of Conservation
North Dakota	Daniel S. Boone, Petroleum Engineer, Amerada Petroleum Corporation with the co-operation of Clarence B. Folsum, Jr., North Dakota Geological Survey
South Dakota	Allen F. Agnew, State Geologist, South Dakota Geological Survey
Nebraska	H. N. Rhodes, Director, Nebraska Oil and Gas Conservation Commission
Kansas	Albert E. Sweeney, Jr., Office of Oil and Gas, United States Department of the Interior
Oklahoma	F. H. Rhees, Vice-President and W. J. Rogers, Sinclair Oil and Gas Company

from known deposits. W. D. Dietzman, M. Carrales, Jr., and C. J. Jirik, *Heavy Crude Oil Reservoirs in the United States*: *A Survey*, U.S. Bureau of Mines Information Circular 8263 (Washington: 1965). It reports that 75 per cent of the heavy oil (gravity of 25 degrees API or less) reservoirs in the U.S. contain, as of January 1, 1964, over 90 billion barrels of oil in place. If deposits with little or no production history are added, the figure rises to over 150 billion barrels. While the report itself does not speculate on what percentage of this oil is recoverable, the press release announcing the study (U.S. Department of the Interior, Press Release, May 3, 1965) estimates that from 30 to 60 billion barrels can be recovered with the application of heat.

Arkansas	Paul D. Torrey, Petroleum Engineer
Alabama	H. Gene White, Petroleum Engineer, State Oil and Gas Board of Alabama
Mississippi	James F. Borthwick, Jr., Chief Engineer, Mississippi State Oil and Gas Board
Louisiana	A. Fred Peterson, Jr., Petroleum Engineer, Louisiana Department of Conservation
Texas	Paul D. Torrey, Petroleum Engineer
New Mexico	Daniel S. Nutter, Chief Engineer, New Mexico Oil Conservation Commission
Colorado	A. J. Jersin, Director, Colorado Oil and Gas Conservation Commission
Wyoming	J. Howard Barnett, Petroleum Consultant and Chairman, Wyoming Secondary Recovery Committee
Montana	James F. Neely, Executive Secretary, Oil and Gas Conservation Commission of Montana
Utah	Robert L. Schmidt, Chief Petroleum Engineer, Utah Oil and Gas Conservation Commission
California	G. B. Shea, Chief, San Francisco Petroleum Research Laboratory, U. S. Bureau of Mines
Alaska	Donald D. Bruce, Chief, Petroleum Branch, Division of Mines and Minerals of Alaska
Other States	Paul D. Torrey, Petroleum Engineer

ABANDONMENT OF THE IOCC SERIES

As of May 1965, the IOCC Committee had no plan for continuing the series of biennial reports. This, we understand, is due to lack of sufficient co-operation from the state regulatory agency of Texas.

It is perhaps significant that in December 1964, the Governors' Special Study Committee of the IOCC which studied the entire array of petroleum conservation problems made the recommendation that:

The Commission [IOCC] should initiate a continuing study of domestic petroleum reserves, the years of supply at projected rates of consumption, and the effects on the domestic industry of increasing consumption of gas and gas liquids, imports, and of new development, whether on private or public lands, onshore and offshore.[8]

[8] *A Study of Conservation of Oil and Gas in the United States, 1964,* Purpose and Nature of Study, Summary and Conclusions, Report of the Governors' Special Study Committee, Interstate Oil Compact Commission (Oklahoma City: 1964), p. 2a.

The report does not spell out what type of "continuing study" is contemplated. Hopefully, it would revive the discontinued IOCC series and work toward refining its concepts and numbers.

The disappearance of the IOCC estimates would create a complete statistical gap in estimating procedures with respect to fluid injection and original oil in place. This raises the question of the importance of having such information. In view of the apparent decline in the rate of discovery of large new reserves, added importance is likely to be attached to the potentialities from secondary recovery and improved recovery practices. We are in no position to judge the quality of the IOCC estimates, but their disappearance is bound to be regarded as a loss by public agencies and private investigators concerned with problems of future availability of energy.

The question will no doubt arise whether the API should be urged to enter these fields of estimation. The API Committee has from time to time considered the feasibility and desirability of estimating oil originally in place, but without action. The subject is one surrounded by difficult technical problems and large gaps of knowledge; and the technical personnel with whom we have discussed the matter have shown no enthusiasm for undertaking such a task. This is not, however, for lack of interest. The API has had a Subcommittee on Recovery Efficiency at work since 1956. In the late stages of preparing a report based on study of several hundred reservoirs, the subcommittee planned, among other things, an examination of percentage recovery of oil in place under various conditions, defining recovery empirically as a function of various parameters, such as rock, type of fluids, environment, and type of drive. Such a report should add greatly to knowledge concerning recovery potentialities, within ranges determined by economic factors.

Knowledge acquired in this painstaking way will not, however, add up to a neat over-all figure of oil in place or of percentage recoverable under present or alternative economic and technical conditions. Designed for other purposes, it is too partial, too detailed, too surrounded by reservations. Nevertheless, it is conceivable that it might serve as the basis for a more generalized picture of the outlook, suitable for the understanding of technically untrained people. After a brief education in the problems of estimation, we can claim no capacity to prescribe the form or content of information which the API might reasonably be asked to provide. In view, however, of the mounting interest of public agencies in such information, it may be supposed that the industry will be taking a fresh look at what it might be able to do.

The virtue of the IOCC estimates is, perhaps, not so much in the numbers themselves, as in calling attention to the extremely limited meaning of the API numbers, and in expressing the need for concepts and numbers with which to discuss the problems of future availability of oil. With the disappearance of the IOCC estimates, the question is what to put in their place to serve this purpose.

5

Company Estimates of Reserves for Company Purposes

The API method of estimating reserves in the aggregate has its roots in the procedures used by individual companies for estimating their own reserves. Companies make such estimates and have for this purpose expert personnel, which is drawn upon to man the committees of the API. At the heart of the API reports lies the fact that every company has a need for reserves estimates of the nature of "proved" reserves, and, while individual companies need not follow an absolutely uniform method of estimating them, the estimates can be adjusted and pooled through the extensive knowledge of co-operating experts. The ideology behind the API estimates therefore reflects ideologies prevalent in the industry.

Data on proved reserves are needed by companies for a number of purposes. In addition, for certain purposes, they also need estimates of the productive possibilities of their properties on a less restrictive and more uncertain basis than is applied to estimating proved reserves. In exploring the "probables" and "possibles" of their properties, separate companies do not, however, apply anything like uniform procedures of estimation. There is also a great difference between companies as to the extent to which, and the regularity with which, they make such evaluations. This is perhaps the basic reason why—although in general they all know that their ultimate production potential greatly exceeds their proved reserves— no attempt is made in API reports to quantify this larger potential in the aggregate. Detailed and comparable data for aggregation do not exist, though expert opinion in the industry is capable of making well-informed guesses. Another barrier to such reporting is, no doubt, the fact that companies do not wish to disclose to one another the estimates they place upon the potential of their own holdings or of areas in which they have a prospective investment interest.

The whole reporting situation can be better understood by examining the company uses of reserves data.

COMPANY USES OF RESERVES ESTIMATES

The individual company estimates its own reserves for a number of different purposes. Such estimates are required for corporate and tax accounting, especially in connection with setting cost depletion, depreciation and amortization, and for income tax calculations. For these purposes something in the nature of "proved" reserves is customary and, indeed, necessary.

Beyond these two formal purposes, the company estimates largely fall under the concept of "evaluation," which may be required for a variety of purposes. Of these, the most important is, no doubt, that of internal audit and management control, especially with respect to the evaluation of the results of past investment and the directions of new investment. For this purpose it appears customary to have a base figure of proved reserves. However, for evaluation purposes it must at times be necessary to have in view a broader concept of reserves than that of proved reserves. The "probables" and "possibles" of the situation have some place in evaluations, as well as the potentialities of possible secondary recovery measures. But it appears to be not merely customary, but informative and necessary, for management purposes to have these figures shown separately from the base figure representing the "proved" concept.

Some types of evaluation call for much more detail than others, as in the case of purchase or sale of producing or prospective properties, or in the arrangement for joint operations or unitization plans calling for the definition of participating interests in the joint operation. In such cases, while proved reserves may provide the point of departure, much more information is required to define the respective interests.

Evaluations are also called for in connection with financing operations, as in reports to banks and securities companies to support loans or securities flotations, though here the interest of the lender is more in assured rates of production during a definite period than in estimated underground inventories. Reserves estimates also have to be supplied to the Securities and Exchange Commission to support requests for permission to float securities; but how far these estimates tally with estimates made for other purposes is impossible to know.

The company interest in estimates of its own proved reserves may be summarized in the following excerpt from an internal company document:

Estimation of hydrocarbon reserves is the base for determining _____Company's underground assets, and establishing the success of our own efforts to remain a dominant factor in domestic energy production and supply. The worth of a clear, moderately detailed, survey of proved reserves to executive officers, as well as complementary departments or groups, becomes clearer when a few of the uses of the report are recognized. Reserves are used to evaluate results of exploration and development expenditures, to channel funds for future expansion of producing operations, for local tax assessments and negotiations, for corporate depletion and depreciation accounts, determination of equities under unitized operations, for stockholder relations, and for analysis by financial institutions.

While this statement runs in terms of proved reserves only, the same company includes in its estimating procedures an estimate of "prospective" (or probable) reserves. It would appear that no company could fail to use this category, however loosely estimated, if it wishes to exercise intelligence in the direction of its developmental funds.

The reasons for company estimates of their own reserves are more fully developed in a study, "Valuation in the Petroleum Industry,"[1] in which the consulting firm of DeGolyer and MacNaughton analyzes the reasons for company estimates of their own reserves and presents a classification and description of reserves categories.

While, as they say, "most of the valuation work in the oil and gas producing industry is concerned with the estimation of recoverable hydrocarbon reserves from developed properties and the appraisal of the future net revenue to be realized from the production and sale of the reserves," there are circumstances which require valuations of undeveloped properties and of undrilled "wildcat" acreage not reducible to quantitative statement in terms of barrels or cubic feet. The estimation of reserves therefore requires techniques applicable to these various circumstances.

They list a number of reasons for which valuations may be needed; some of these are:

---

[1] Reprinted from *Oil and Gas Taxes* (New York: Prentice-Hall, 1962), pp. 2041ff.

1. To establish the collateral value of properties in loan financing and in purchases involving production payment contracts.

2. To determine the market value of a property in connection with possible purchase or sale.

3. To satisfy the Securities and Exchange Commission and the Department of Justice in connection with corporate financing, corporate mergers or dissolutions.

4. To determine the remaining oil and gas reserves as a factor in figuring depreciation and depletion for tax purposes.

5. To determine the feasibility of secondary recovery projects and other special recovery projects.

After noting a few other minor reasons, they go on to say:

> Perhaps the most important function of valuation is to provide management with a determination of its own reserves of oil and gas. Petroleum reserves usually constitute the only source of income for an integrated company. . . . All long-range planning for a producing company should stem from an analysis of estimated net income and future capital requirements needed to maintain current producing properties, to provide for development drilling, and to search for additional hydrocarbon reserves.

They continue with a statement of the necessity of estimates not only of proved reserves, but of reserves in lesser categories of certainty, as follows:

> For purposes of establishing collateral, depletion allowance, and for most tax purposes and other financial uses, only proved reserves are of material consequence. Probable and possible reserves are important because they provide an incentive for spending additional capital in exploration. . . . Successful conversion of these reserves to the proved category together with the development of secondary reserves may, to a large extent, provide the reserves which replace production and enable operators to maintain or enlarge their position in the petroleum industry.

This brief survey of company uses of reserves data indicates that, while "proved" estimates serve a number of specific purposes, supplementary estimates are required for certain other purposes.

SYSTEMS OF CLASSIFICATION FOR COMPANY USE

An examination of the procedures used by various companies for estimating their own reserves shows that there is no uniformity from company to company. This entails a number of disadvantages and has led to proposals for greater uniformity.

The Lahee Classification

A pioneer proponent for more uniform methods of classifying reserves is Frederic H. Lahee, long prominent in the activities of the API Committee on Petroleum Reserves. His proposed classification is as follows:[2]

A. Proved reserves, which may be:
    1. Drilled, or
    2. Undrilled.
B. Probable reserves, which may be:
    1. In the still undeveloped parts of a pool;
    2. Obtained by installation and operation of secondary recovery methods;[3]
    3. In tested sands (reservoir rocks), now cased off, above the producing pool.[3]
C. Possible reserves:
    1. On structures or in environments now producing, either
        a. In undiscovered pools below the producing pool, or
        b. In undiscovered pools laterally outside the limits of the producing pool.
    2. On structures or in environments not now producing, which are similar to other producing structures or environments in the region.
D. Hypothetical reserves.

The DeGolyer-MacNaughton Classification

In the study referred to above, DeGolyer and MacNaughton present their own method of classification as follows:

---

[2] "The Terminology of Petroleum Reserves," *Proceedings of the Fourth World Petroleum Congress.* Section II/H. (Rome: 1955).

[3] It has been commented that this item may include a substantial amount of proved reserves.

The primary ingredient of the appraisal of oil and gas properties is an estimate of the recoverable reserves of oil and gas. Such an estimate, to be reliable, must be based on a comprehensive and detailed study of the basic facts involved in the production of oil and gas from the properties being appraised. These basic facts are interpreted by the valuation engineer with the use of a broad background in the physical sciences combined with adequate experience and good engineering judgment. The result then is an estimate of the reserves of oil and gas which are recoverable from these properties, the estimate being arranged in such a manner that its several parts can be properly related to the revenue that will accrue from it. For these purposes, reserves of oil and gas can be divided into the following categories:

(1) *Primary Reserves*—Estimated future commercial production recoverable by normal or primary methods as a result of energy inherent in the reservoir.

(a) Proved—Reserves which have been proved to a high degree of certainty by reason of actual completion, successful testing, or in certain cases by adequate core analyses, and which are defined areally by reasonable geological interpretation of structure and known continuity of oil- or gas-saturated reservoir material.

1. Developed—Including those classified as producing and those classified as non-producing. Producing Reserves for this purpose are those to be produced by existing wells in present completion intervals with normal operating methods and expenses. Non-Producing Reserves for this purpose are those behind the casing or at minor depths below the producing zones which are considered proved for production from other wells in the field, by successful production tests, or in some cases, by adequate core analyses from the particular zones.
2. Undeveloped—Reserves which are considered proved for production by reasonable geologic interpretation of adequate sub-surface control in reservoirs producing or proved by other wells but which are not recoverable from present wells.

(b) Probable—These reserves, susceptible of being proved, are defined by less direct well control but are based upon evidence of producible gas or oil within the limits of a structure or reservoir above known or inferred water saturation.

(c) Possible—These reserves are less well defined by structural control than probable reserves and may be based largely on electric log interpretation and widespread evidence of crude oil or gas saturation. They also may include extensions of proved or probable reserve areas where so indicated by geophysical or geological studies.

(2) *Secondary Reserves*—Estimated future commercial production which will be recovered in addition to the primary reserves as a result of pressure maintenance, water flooding, or other secondary methods.

(a) Proved—Reserves which have been proved to a high degree of certainty in reservoirs where secondary recovery methods are already in normal operation or have been proved economic by pilot operations, and in certain cases where successful secondary recovery operations are taking place in similar nearby reservoirs producing from the same formation.

(b) Probable—Reserves in reservoirs which appear to be suitable for secondary recovery operations but for which inadequate data are available to determine the extent and increased recovery to be expected.

(c) Possible—Reserves in reservoirs which appear to be for secondary recovery operations but for which very few data are available.

The Arps Classification

A somewhat similar classification, with careful definitions and extended to cover secondary recovery, has recently been proposed by J. J. Arps, a prominent petroleum engineer and evaluation expert. His work is shown in Table 8.

THE NEED FOR UNIFORM CLASSIFICATION

While differing in descriptive detail, all the descriptions of reserves estimation procedures we have quoted (Lahee, Arps, DeGolyer-MacNaughton) agree in principle upon certain fundamentals: (1) that for company purposes estimates of reserves should include the categories of proved, probable, and possible reserves, and (2) that these categories have to be applied separately to the categories of primary and secondary reserves.

Compared with this degree of detail required for intelligent company planning and investment, the only data aggregated into national estimates by the API are the proved primary reserves, technically defined, plus small allowance for secondary reserves where installations have been successfully applied.

As against the efforts of people like Mr. Lahee and Mr. Arps to introduce closely defined orders of probability into the estimates of reserves,

TABLE 8. CLASSIFICATION OF PETROLEUM RESERVES

Source of reservoir energy	Degree of proof	Development status	Producing status
Primary Reserves recoverable commercially at current prices and costs, by conventional methods and equipment, as a result of natural energy inherent in the reservoir.	**Proved** Primary reserves which are considered proved to a high degree of certainty by actual production from the reservoir at commercial rates, or in certain cases by successful well test(s) in conjunction with favorable and reliable core analysis data or quantitative log interpretation. This indirect evidence is valid only if no substantial drainage has occurred since these well(s) were drilled or tested.	**Developed** Proved reserves recoverable through existing wells. **Undeveloped** Proved reserves under undeveloped spacing units which are so close and so related to developed spacing units that they may be assumed with confidence to become commercially productive when drilled.	**Producing** Developed reserves to be produced by existing wells from completion interval(s) open to production. **Nonproducing** Developed reserves to be produced from existing wells but which are now behind the casing or at minor depths below the bottom of the hole. The cost of opening up such reserves should be relatively minor.
	Probable Primary reserves behind the casing of existing wells or within the known geological limits of a productive reservoir, which are inferred from limited evidence of commercially producible oil or gas, but where the evidence is insufficient to qualify under the "proved" definition.		
	Possible Primary reserves whose existence may be inferred from geological considerations but where available data will not support a higher classification.		

TABLE 8. CLASSIFICATION OF PETROLEUM RESERVES—Cont'd.

Source of reservoir energy	Degree of proof	Development status	Producing status
Secondary Reserves recoverable commercially at current prices and costs, in addition to the primary reserves, as a result of supplementing by artificial means the natural energy inherent in the reservoir; sometimes in conjunction with a change in the physical characteristics of the reservoir fluids.			Producing Developed reserves to be produced by existing wells from that portion of a reservoir subjected to full scale secondary operations.
	Proved Secondary reserves which are considered proved to a high degree of certainty by a successful pilot operation or by satisfactory performance of full scale secondary operation in the same reservoir, or in certain cases in a similar nearby reservoir producing from the same formation.	Developed Proved reserves, recoverable through existing wells from a reservoir where successful secondary operations are in progress.	Nonproducing Developed reserves to be produced by existing wells upon enlargement of existing secondary operations, provided the cost of such expansion is relatively minor.
		Undeveloped Proved reserves which may be assumed with confidence to be produced upon the installation of a secondary recovery project and/or by the drilling of additional wells.	
	Probable Secondary reserves which are inferred from past production performance or core, log, or reservoir data, but where the reservoir itself has not been subjected to secondary operations.		
	Possible Secondary reserves from reservoirs which appear to be suited for secondary operations but where available data will not support a higher classification.		

SOURCE: J. J. Arps in Society of Petroleum Engineers, Dallas Section, *1962 Symposium on Petroleum Economics and Valuation*, facing p. 12.

W. S. Eggleston, in the Symposium referred to above,[4] took a firm conservative position, saying: "Neither should the 'indicated', 'possible', 'inferred', or so-called 'geological reserves' be dignified by the term 'reserves'." And he quotes another writer with approval: "A tract of land is either proved or it is not proved; and if it is not proved then no circumstance warrants crediting it with reserves." Commenting on this, Mr. Arps said:

> I agree with him that for most of the purposes listed in his paper only "proved" reserves should be considered, while the use of terms such as "probable" and "possible" should generally be avoided and certainly should be confined to the descriptive portion of an appraisal report. However, in evaluating the results of an exploration program, or when considering the geologic potential of a given basin or area, there is occasionally a need for terminology which is somewhat less severe than the "proved" definition. The term "probable" and "possible" may fill that requirement, provided it is clearly understood what their limitations are.

The concern felt by many at the disorderly state of reserves concepts and terminology was expressed in the same Symposium by W. W. Wilson, a banker and engineer, as follows:

> Most, if not all reservoir studies are made for the purpose of equating a given course of action with ultimate profit, and therefore come under the general classification of evaluations. Reports of this type may be made for a variety of specific purposes, but ultimately they are used by management at some level for the solution of problems or the formulation of policy. The managements of most companies have established specific policies to guide their engineering staffs, in order to assure uniformity of definitions and nomenclature. So long as these reports are restricted to internal use this approach is workable and presents few difficulties. Serious problems will occur, however, whenever comparisons are to be made of reserve summaries or evaluations prepared by the technical staffs of two or more companies, each having their own ground rules. This is the one main reason why many companies are unwilling to publish reserves data in reports to stockholders, regulatory agencies or security analysts.
>   Most consulting engineers, when first starting their independent

---

[4] *1962 Symposium on Petroleum Economics and Valuation.* Society of Petroleum Engineers of the American Institute of Mining and Metallurgical Engineers (AIME), Dallas, 1962.

practice, tend to follow the procedures used by their former employers. Because of the wide diversity of procedures followed by consultants with different backgrounds of experiences and training, serious problems are created for lending institutions, security analysts, and other persons who use the reports they prepare. . . .

During the last several years, the writer has made an informal review of the procedures used by producing companies, consultants, and bank engineers, and has reviewed the available literature on evaluation procedures in an attempt to define the scope of this problem. The results are both enlightening and confusing. There seems to be little uniformity in definitions of various types of hydrocarbon reserves, methods of classifying such reserves in accordance with their degree of certainty or producibility, and the nomenclature used in evaluation reports.

There appear to be very few published definitions specifically defining crude oil, natural gas, condensate, and natural gas liquids. Such definitions as were found do not agree in all cases, and the degree of difference is sufficiently large that confusion is unavoidable. Some companies follow the practice of classifying as proven reserves only the hydrocarbons recoverable from either existing wells or completions, and give no credit to undeveloped reserves in any category. Certain other companies employ what amounts to an accounting system for reserves records, and transfer reserves from one classification to another as specific locations are drilled or as wells are recompleted. There are, of course, many other systems somewhere between these two extremes. There appears to be no uniform manner of accounting for the various categories of undeveloped secondary recovery or pressure maintenance reserves.

In most instances, consulting engineers follow the dictates of their clients with regard to classification of proven or speculative reserves, but in the absence of specific instructions each seems to follow his own set pattern. There is little uniformity with respect to definitions, or classification of reserves between individual consultants. The problems which this situation creates are compounded whenever it is necessary to composite the results of two different consultants, each of which has evaluated a part of a total package of producing properties. Every bank engineer and consultant is aware of many instances when engineers have been criticized severely because of their failure to agree within reasonable limits in evaluating specific properties. No doubt much of this ill will has been created unnecessarily and is a direct result of confusion with respect to definitions and reserve classification.[5]

[5] *Ibid.*

Following the interest stirred by the Symposium, and in response to requests by industry groups, the President of the Society of Petroleum Engineers (SPE) took the initiative in setting up a "Special Committee to Develop Definitions of Classifications of Proved Petroleum Reserves Necessary for Property Evaluation" (or, for short, "SPE Committee on Definitions for Proved Reserves"). The Committee, as constituted in 1963 under the chairmanship of J. J. Arps included representatives of crude oil producers, natural gas producers, consulting engineers, bankers and insurance companies. The chairman of the API Reserves Committee was included, but the American Gas Association (AGA) Committee did not undertake direct participation. A strong activating force in the setting up of the SPE Committee appears to have been the financial institutions, for whom the lack of uniformity was a confusing factor.

The instructions to the Committee were to evolve definitions of classifications of proved petroleum reserves necessary for property evaluations, as distinguished from the estimation of regional or nation-wide reserves. Among the aspects to be considered were: (1) the source of reservoir energy, (2) the degree of proof required to attain approved status, (3) the development status, and (4) the producing status. By reference to Mr. Arps's table (Table 8), it will be seen that these are his basic classifications of factors.

Since the Committee at the present time of writing (May 1965) has not produced a final report, it is not possible to state the final outcome of this effort. We can, however, indicate the general direction of its purpose. In the first place, it is concerned only with the definition of proved reserves. It is attempting to effect a harmonious adjustment between the API definition and the more detailed categories required for evaluation of individual properties. The presumption is that it will suggest certain clarifying changes in the API definition, and will then propose more detailed classifications within the framework of the API definition.[6] Such greater detail would appear in such categories as whether the reserves are developed or undeveloped, and if developed, whether they were producing, or nonproducing. While the Committee's work is confined to defining only proved reserves for company evaluation purposes, the fact that it has a real subject-matter appears in the fact that it turned up no less than a dozen different published definitions in actual use. The establishment of uniform standards is deemed by the Committee to represent a service to the industry comparable, for example, to the standards of well classification effected by the American Association of Petroleum Geologists.

[6] The same may be said of the AGA definition, if this body ultimately co-operates in the work of the Committee.

For purposes of the present study, the most noteworthy point about the work of the Committee is what it does *not* include. It does not include any attempt to reduce to uniform definition the categories of "probable" and "possible" reserves. As we saw earlier, systems of company evaluation of properties have some place for the use of these categories; and it may be that early ideas of the role of the SPE Committee envisaged some effort to standardize usage. It was in any case decided that no attempt should be made to take this matter out of the realm of judgment and practice of each individual company.

It is not surprising that this decision should be reached in view of the strong judgment factor which necessarily enters into all "probability" estimates, the different experience of companies, and the different methods of evaluation they have used. The decision does, however, have a bearing upon proposals that the API, in addition to its estimates of proved reserves, should provide supplementary estimates based upon orders of probability. There is no present statistical basis for aggregating company estimates of their "probable" reserves. Even if there were, it seems highly probable that companies would not care to expose to view their own estimation of the productive potentialities of their properties. Given the competitive structure of the industry and the non-uniform character of evaluation practices, the voluntary expansion of reserves information in this direction from industry sources seems highly improbable.

PROJECTIONS OF CURRENT DISCOVERIES FOR COMPANY USE

We now turn to a special aspect of the conventional basis upon which reserves from new discoveries are estimated in the "proved" category of both company evaluations and API reports. In the case of the API, it was noted earlier that the first year estimate is made upon the basis of a single well, or a few wells, and confined to the recovery from a very narrow geographical area. Everyone "knows" that in the course of time this figure will normally be upped several times over by "extensions and revisions," not necessarily for any single discovery but certainly for numerous discoveries taken together. The question arises whether this potential is capable of current statistical estimation. The root of the problem lies at the level of company estimating procedures.

One of the responsibilities of a company management is to evaluate the results of its exploration program. This means, in effect, balancing the oil "found" with capital expenditures over time. If year by year a company simply records the "proved" reserves from exploration on a conventional basis, it will have to wait quite a few years before it has any clear notion of what it has found in the recent past. This system of recording results is apparently what takes place in most companies, though no doubt managements discuss the possibilities of future recovery from recent discoveries with their geologists and engineers. The element of guesswork is necessarily large unless systematic analytical procedure is applied. Some statistical ingenuity has been put to reducing the amount of guesswork in order to achieve greater certainty in the current evaluation of future results from new discoveries.

Possibilities of this type may be illustrated by an analytical technique devised by J. R. Arrington, formerly chief economist of the Carter Oil Company and now with the Humble Oil Company.[7] Mr. Arrington presents the problems as follows:

> Every company searching for crude-oil reserves is constantly faced with the problem of evaluating its exploration program. The efficiency of a program can best be measured by relating exploration expeditures to the barrels of oil found. Usually the dollars can be determined or estimated, but the barrels of oil discovered by these expenditures are difficult to establish. The industry does not immediately know the size of its discoveries; there is a definite time lag before the adequate valuation of reserves can be made. Thus the true results of current exploration are not known. Unfortunately, oil men must make decisions involving large financial risks before they know all of the facts. The techniques for adjusting reserves described in this article are aimed at improving the evaluation of exploration programs much earlier.
>
> The most widely accepted published estimates of oil found by current wildcats are the initial estimates of discoveries by the API. These data have limitations. Initially, the estimates are conservative but more important, it is not possible to follow changes in a given year's discoveries or to trace their growth. Each year subsequent changes in the estimates of a particular year's reserves are lumped with discoveries of other years as they are extended and revised.
>
> The industry needs a system which will:
> 1. Isolate reserve estimates, grouping them by year of discovery.
> 2. Allocate each reserve change back to the year of discovery.
> 3. Adjust current estimates for future revisions.

[7] "Predicting the Size of Crude Oil Reserves, etc." *Oil and Gas Journal*, February 29, 1960.

Some elements of the plan may be stated in Mr. Arrington's own words:

Definition. Before developing the method, the following terms should be explained (Reference will be made to three types of reserve estimates):

1. The initial estimate—this is the first estimate of reserve for a new reservoir or the estimate as recorded at the end of the discovery year.

2. Revised estimate—this is the estimate at a specific point in time after the initial estimate. It includes reserves changes year by year after the discovery year.

3. The probable final estimate—this is a statistically developed estimate which anticipates future revisions, and is the estimate to be described in this paper. Revised estimates approach the probable final estimate with each year of revision.

Description of the method. To develop a statistical method of predicting future revisions, the growth of discoveries must be followed from year to year; this requires that reserve estimates be grouped by year of discovery. Only a few companies maintain such records for all fields discovered by the industry.

If reserves are maintained by year of discovery for a number of years, the rate of growth of each year's discoveries can be followed.

Without going into the details of the plan, the eventual result will be to develop a factor which can be applied to current estimates of discoveries, adjusting them for anticipated future revisions. The system would initially apply only to primary reserves; but it could be supplemented by later results achieved through fluid injection. It could not be introduced as a predictive instrument at once, except by companies which have grouped their reserves by year of discovery, since several years' experience would be required to develop a suitable factor.

In Mr. Arrington's words, "The most important advantage of having probable final estimates is the ability to measure results of the current exploration programs. . . . This is a valuable tool for management in helping them to learn how facts which might not otherwise be available for 5 to 10 years."

From inquiries we have made, we learn that at least a few companies keep records on a plan comparable to the one described by Mr. Arrington, so that all later additions to reserves can be ascribed, pool by pool, back to the original year of discovery. If such a system were generally adopted and the results made available, the potential results of current discoveries could be statistically projected and the trend of exploratory results traced for the whole industry. Since there is no reason to expect any such development in the calculable future, aggregated statistical results of this sort are unlikely to be forthcoming. Nevertheless, since API "extensions and revisions" have in the first instance to be based on field-by-field data, there appears to be no insurmountable obstacle to achieving something roughly comparable for the industry as a whole through expanded analysis based on API data.

INTERNAL OBSTACLES TO EXPANDED ESTIMATION

In concluding this account of company interest in reserves, it is necessary to note the existence among managements of a variety of attitudes towards reserves estimates. To quote an officer of a large company, "Inside companies there is as much confusion concerning reserves as there is in the outside world." Certainly, there is no uniformity in their ways of estimating their own reserves or in the manner of thinking of top management about them.

The most extreme position we have encountered is the statement by a high oil company official that "knowledge of its own reserves is of no use to a company," intended no doubt as an exaggeration to give force to the further statement that "the company's interest is in production." It is of course necessary to have reserves in order to produce. But the cash revenues each year represent production from reserves revealed by outlays well in the past. The primary problem of management is what to do with these revenues. When they are ploughed back in the further search for oil, it is an act of faith based on the hope that, one year with another, they will reveal enough new reserves to keep the company permanently in the oil business. The precise size of the current reserves, whether larger or smaller and whether more or less certain, does not change the character of management investment decisions.

The rate of accrual of new reserves is, of course, a datum of interest to companies, as showing whether they are running ahead of, or behind, the production game. But even for this purpose, most companies appear to be content with a figure for proved reserves from which they can, if they wish, arrive at some estimate of "probable" content by a rule of thumb based on their past experience and that of the industry. Most companies, it appears, utilize more elaborate evaluations of their properties only for special occasions, such as purchase and sale of properties, mergers, and transfer of leases.

While some companies, or at least some of their technical staff, certainly take a more active interest in reserves on a longer-run view, it appears to be the case that few companies have systematic procedures for evaluating their "probable" and "possible" reserves according to orders of relative certainty or uncertainty. On the other hand, most companies must and do make judgments about their probable and possible reserves, and also make decisions based on these judgments. The analytical exercises for systematic appraisal would, we gather, be considered by most managements as a waste of effort and money. The question they have to face is whether the added cost of such analysis each year would be justified by the degree of improvement in their planning and investment. So long as the answer is largely in the negative, the industry cannot provide the statistical data from which to aggregate estimates of the amounts of oil to be regarded as recoverable from known fields within categories of relative certainty or uncertainty. Expert opinion within the industry could do some shrewd guessing on this point if the experts wished to pool their knowledge. But even such "unofficial" conjecturing, it must be recognized, would have to proceed within strictly defined economic and technological assumptions.

6

An Alternative Approach: Calculating Reserves from Producing Capacity: The A. D. Zapp Study

The two series of estimates described earlier, the API and the Interstate Oil Compact Commission (IOCC), provide the point of departure for most other estimates of the future availability of oil under less restrictive assumptions. At later points we shall review some special studies of this sort. Below we introduce a proposed method of estimating reserves which rejects the API approach and ignores the IOCC approach.

The proposed method is presented in *Future Petroleum Producing Capacity in the United States* by the late A. D. Zapp, a study issued by the U.S. Geological Survey.[1] This study is one of method and does not attempt quantitative estimation, since the data do not now exist to which to apply it. Mr. Zapp's proposed procedure is (1) to start from the measured producing capacity of existing wells, (2) to estimate the reserves (probable future recovery) underlying these wells by the use of decline ratios, and (3) to project reasonable expectations of future additions to supply.

The abstract of this study opens as follows:

Prediction of future petroleum producing capacity should be based on the statistical record of the past, interpreted in the light of the known trends and circumstances that are not amenable to statistical expression. For a dynamic industry such as the petroleum industry,

---

[1] *Geological Survey Bulletin 1142-H* (Washington: Government Printing Office, 1962).

the most recent statistical record should be the most meaningful as to the future.

The following words introduce the main body of the paper:

Any forecast of the future petroleum producing capacity of the United States obviously must be based on available measurements or quantitative estimates pertaining to petroleum, and there are several possible approaches to the problem, depending on which set of measurements or estimates is chosen as fundamental. Recent published predictions as to the future producing capacity of the United States have been based primarily on either the yearly estimates of proved reserves prepared by committees of the American Petroleum Institute and the American Gas Association or estimates of the total amount of ultimately recoverable petroleum, including the undiscovered quantity. A third possible basis for prediction is the recent record of increase in producing capacity, as shown by a series of recent surveys sponsored by the National Petroleum Council and the Petroleum Administration for Defense. Careful consideration was given to the nature and intrinsic accuracy of each type of data or estimate and the historical trends affecting them. From this study, it was concluded that the recent record of development of producing capacity is the most meaningful as to what may be expected in the foreseeable future. The reasoning that led to this choice is summarized in this paper.[2]

In his study Mr. Zapp comes quickly to the conclusion that the proved reserves estimates of the API provide no basis for predicting future availability of oil. Their primary defect for this purpose is that the current statistics of additions contain a large "lag element." Current additions to reserves are in substantial degree revisions of the reserves earlier attributed to fields or pools discovered in the more or less remote past. ". . . because of the lag element in reserve estimations, recent trends based on proved reserve statistics are deceptive."[3] Referring, for example, to the history of the discovery of "giant" fields (capable of ultimate production exceeding 100 million barrels), he notes that "at the end of 1948, only 12 of the fields discovered during the preceding 10 years were recognized as 'giants' —10 years later the count had grown to 32. . . ."[4] Moreover, ". . . the upward revision of reserve estimates results not only from additional

---

[2] *Ibid.*, p. H-3.
[3] *Ibid.*, p. H-12.
[4] *Ibid.*, p. H-12.

drilling (development drilling) that increases the known extent of a pool, but frequently results from upward revision of estimates for previously drilled parts of pools."[5]

Consequently, because of this lag element—the current dating of additions to reserves from pools of earlier discovery—the API figures provide no clue to recent statistical trends. As his study notes, "the yearly estimates of proved reserves do not purport to be a measure of the discovery rate, and should not be interpreted as such."[6] In spite of this, some writers on the industry have found it:

> tempting to interpret the quantitative expressions of total "new oil" added during the year . . . as a measure of the rate of development of new supply, and to draw conclusions as to future supply by projecting trends based on these statistics. Thus, Pogue and Hill . . . have plotted and projected the statistics directly, and have called them "annual discovery rate." Similarly, Davis . . . has divided the yearly totals . . . by the total footage drilled yearly to derive a statistical element termed "drilling return" expressed in barrels per foot drilled, and has based conclusions on projected trends in the "drilling return."[7]

There are still other misuses of the statistics. "The proved-reserve estimates have also commonly been misinterpreted as reflecting the true rate of increase in the quantity of petroleum producible from existing wells without change in production practice, or as reflecting the full quantity that is economically producible from existing wells."[8]

Finding the API statistics of proved reserves incapable either (1) of supporting a reasonable estimate of recovery from wells already drilled, or (2) of projecting the recoverable content of known pools through later development, or (3) of establishing trend relations between past and probable future rates of discovery, Mr. Zapp turns his attention to the kinds of data and the methods of analysis which would throw light on these matters.

His principal clue as to data and method is the series of reports by the National Petroleum Council (NPC) on the producing capacity of the industry, supplemented by the 1945 report of the Petroleum Administration for War (PAW) which dated later "extensions" and "revisions" back

---

[5] *Ibid.*, p. H-12.
[6] *Ibid.*, p. H-15.
[7] *Ibid.*, p. H-15.
[8] *Ibid.*, p. H-15.

to the year of discovery of the field or pool. At the time of writing, he did not have available the 1960 report of the NPC (discussed in Chapter 3 above) which not only updates productive capacity to 1960, but also updates the 1945 report in allocating back to the year of discovery. This lack of access to later data, nevertheless, does nothing to impair his thinking on analytical method.

The basic advantages of starting from estimates of producing capacity as an approach to measuring increases in reserves are, according to the Zapp study, that producing capacity can be measured with relative accuracy and that "estimates of producing capacity are current; that is, they are not subject to the 'lag element'."[9] The bridge between producing capacity and reserves of drilled wells is provided by the "decline-curve" principle. Strictly defined, this principle is that ". . . the rate of production of an individual oil well, if continuously produced at the maximum possible rate and without change in production practice declines steadily with time."[10] A decline curve is shown by the rate of production plotted over time. Normally, the curve declines rapidly during the early period and then at a diminishing rate over the life of the well. Improved methods of production slow down the rate of decline. The introduction of secondary recovery methods may raise the curve, followed by a second cycle of decline.

The concept of Maximum Efficient Rate of Production (M.E.R.)[11] under recent conservation practices has somewhat changed the meaning of "producing capacity," to which Mr. Zapp gives the following definition: "(1) the M.E.R.'s of those wells operated under the M.E.R. concept, and (2) the maximum producing rates of wells that need no restriction and of those not operated under good engineering practice. It is considerably less than the rate of production would be if all wells were 'opened wide.' The effect of restricted early production under the M.E.R. concept is to increase the ratio of recoverable reserves to current producing capacity."[12] Mr. Zapp's study gives the impression that regulatory agencies use the M.E.R. concept as the basis for restricting production, an erroneous idea upon which we will comment later.

[9] *Ibid.*, p. H-15.
[10] *Ibid.*, p. H-4.
[11] The M.E.R. concept itself is not in practice uniformly defined. It is sometimes defined strictly in physical terms and determined by a formula with such variables as reservoir pressure, permeability, porosity and the like. It necessarily includes, however, an economic factor concerned with anticipated costs and revenues over the life of a reservoir. See the discussion in Stuart E. Buckley (ed.), *Petroleum Conservation* (New York: American Institute of Mining and Metallurgical Engineers, 1951), pp. 151ff.
[12] *Geological Survey Bulletin 1142-H, op. cit.*, p. H-7.

Turning from the technical definitional analysis, which we have merely hinted at, Mr. Zapp reviews the series of estimates of producing capacity prepared under the auspices of the National Petroleum Council and the Petroleum Administration for Defense from 1946 to 1957. He considers these estimates to have a considerable degree of accuracy. The growth of crude oil producing capacity was from 4.7 million barrels per day as of January 1, 1946 to 9.9 as of January 1, 1957. (The later comparable figure as of January 1, 1960 is 10.6 million barrels per day.)

After asserting the principle that "with the same production practice and in similar reservoirs, twice the producing capacity means approximately twice the recoverable reserves"[13]—an assertion which needs clarification— Mr. Zapp makes the following calculation: Crude oil producing capacity was 6.7 million barrels per day in January 1951 and 9.9 in January 1957— an increase of 47 per cent. By way of comparison, the API estimates of proved reserves increased by only about 20 per cent in the same period. (As of January 1, 1960, the percentages would be 58 and 25, respectively.) This comparison is directed to criticism, not of the absolute understatement of reserves by the API, but to its understatement of *the rate of growth* of reserves.

Even a calculation of the rate of growth of reserves by reference to the rate of growth of producing capacity definitely understates the rate of growth of reserves, in Mr. Zapp's opinion, since improvements in production technology are increasing the ratio of reserves to producing capacity.

One difficulty in translating producing capacity into absolute reserves figures is, of course, that the exact shape of the average decline curve, if all wells were produced at M.E.R. or capacity, is not known. As a hypothetical exercise, Mr. Zapp starts with producing capacity of 9.9 million barrels per day and applies an assumed constant-percentage decline of 7.5 per cent yearly. The result would be 45 billion barrels of producible reserves underlying existing wells.

The statistical prerequisite for arriving at "real" figures would be much better information than now exists about actual decline rates. Mr. Zapp in his study, therefore, uses his analysis only for supporting the qualitative conclusion that "the American Petroleum Institute estimates reflect only part of the total quantity producible from existing wells."[14] There are "large quantities that history has taught us will be added to the proved-reserve estimates through 'revisions and extensions' even if there were no

[13] *Ibid.*, p. H-16.
[14] *Ibid.*, p. H-18.

further discoveries."[15] But there are at present "no real measures of all the components of petroleum-already-found."[16]

This conclusion will convey no news to those who are properly informed about the nature of the API estimates. However, Mr. Zapp's analysis serves two useful purposes. First, it documents the failure of the API estimates to support any conclusions concerning the future availability of oil from existing fields, which, it needs to be repeated, is not their purpose. And, more important, it opens a constructive attack upon the problem of how more comprehensive estimates of future availability may be arrived at.

Turning from the problem of estimating the reserves underlying existing wells, Mr. Zapp proceeds to examine the question of how to estimate future accretions of reserves from wells not yet drilled and pools not yet discovered. He reviews studies of the natural geological incidences of oil deposits, the efforts to estimate "ultimate reserves" (in the sense of total recoverable oil in place), and the effect of different estimations of recoverability, with due regard to future technological advance.

Setting aside that highly speculative realm of investigation, he reverts to a down-to-earth question of what kind of statistical data would permit relatively firm estimates of future accruals to reserves from new discoveries, as well as accruals from already discovered fields. A part of the evidence comes from extrapolation of recent exploratory success experience, the uncertain factors being the continuance of this trend and the continuance of economic factors to sustain the necessary incentives. Another part of the evidence is continuous updating of the data on producing capacity, as a guide to what may be expected in the future.

His specific statistical recommendations stem from the principle that "predictions of future trends are best guided by historical trends, and for a dynamic industry such as the petroleum industry, the most recent pertinent statistics should be the most meaningful in regard to the immediate future . . . the estimates of producing capacity are intrinsically the most accurate and most responsive to recent trends . . . the recent record of increase in producing capacity is the most reliable guide as to what may be expected in the foreseeable future."[17]

Extending the same principle, "Factors which could control future producing capacity, and which can best be appraised from a geologic viewpoint are (a) the existence of undiscovered petroleum accumulations sufficient to meet the requirements of expectable increase in production

---

[15] *Ibid.*, p. H-21.
[16] *Ibid.*, p. H-22.
[17] *Ibid.*, p. H-27.

and producing capacity and (b) the ability of the petroleum industry to discover and develop these accumulations at the rate required and under permissive economic conditions."[18]

The basic statistical problems are then (1) data showing the trend of producing capacity which can be extrapolated under various assumptions and (2) data showing the amount of exploratory effort and its results in added producing capacity which can be extrapolated under various assumptions. According to Mr. Zapp's thinking, these two types of data can be used to cross the bridge from producing capacity, present and projected, to estimates of reserves, present and projected, on the decline-curve principle—a point upon which he encounters some professional skepticism.

In this context, Mr. Zapp's proposal is a specific statistical innovation: yearly surveys reporting separately

    (a) the capacity of wells that existed at the time of the previous survey, and

    (b) the capacity of new wells completed during the year.

"Such data would show the average rate of decline in the capacity of older wells and would permit analysis of the amount of new capacity generated in relation to the drilling effort expended."[19]

Mr. Zapp refers, more or less in passing, to the statistical technique of crediting additions to proved reserves through "extensions" and "revisions" back to the year of discovery of the field or pool, in order to establish trends in discovery and provide some expectable magnitude of accruals to reserves in the future from the more recently discovered fields. This type of analysis cuts across the producing capacity analysis but could be collated with it in a fully sophisticated analysis of the prospects for availability of oil in the calculable future. It is to be noted that at the time of writing Mr. Zapp did not have access to the latest NPC report, reviewed in Chapter 3 above. Given his then current interest in trend analysis, the NPC Report might have suggested to him some further lines of analysis complementary to his own.

The public interest in data on petroleum reserves lies precisely in the area where Mr. Zapp centers his analysis: namely, the discovery of trends which can be projected to predict within reasonable limits the future availability of oil. Mr. Zapp discards the API proved reserves series as unsuitable for this purpose; and deals lightly with efforts to estimate "ultimate" reserves. The significance of his method, if put into effect with

---

[18] *Ibid.*, pp. H-29, 30.
[19] *Ibid.*, p. H-30.

relevant data, would apply at any given time to the relatively short-run future, say ten to twenty years, but with a progressive extension of the time perspective.

An advantage claimed by Mr. Zapp for his proposal is that, as to producing capacity, it rests on data sources already established, by regions and nationally, and it would involve only two innovations in procedure: (1) reporting data on producing capacity annually instead of at longer intervals, and (2) separating the capacity of old wells from that of wells opened in the current year. Average decline factors, required to translate capacity into reserves, would have to be calculated from future experience. The critical question is the feasibility of estimating capacity and establishing decline factors accurately and unambiguously under present regulatory practices and the varying circumstances of different fields and regions.

The problem of estimating reserves and future availability of oil would not, of course, be taken care of wholly by such statistical projections. Changing technology and conservation practices and economic factors would enter in. Mr. Zapp's argument, as presented in his study, is that his method would provide a sound point of departure for estimation of future availability, something that is now not provided by the API or any other existing statistical series.

Questions of several different sorts arise in connection with Mr. Zapp's proposal, of which perhaps the central question is whether the data could be provided to implement it. Certainly, they do not now exist, and we have encountered skepticism in expert circles concerning the possibility. In connection with existing wells, the crucial relation is between producing capacity and decline factor. If all reservoirs were produced on the M.E.R. principle, there would perhaps be relatively little difficulty in establishing this relation with respect to individual reservoirs and in the aggregate, and in establishing a statistical trend in these relations over time. The fact is, however, that, under the present system of proration and production control by state agencies, the M.E.R. producing capacity of many reservoirs is not known and the decline factors are thoroughly obscured by production at rates unrelated to the M.E.R.

Most states which allocate production to market demand utilize a depth-acreage formula to determine relative production rates among wells in the state. For example, a 5,000–6,000 foot well on 40-acre spacing onshore in Louisiana has an allowable of 159 barrels per day if it has no market demand restriction. A similar depth well on 80-acre spacing has a 239 barrel daily allowable. At 9,000–10,000 foot depths a 40-acre well has a 274 barrel daily allowable compared to 411 barrels daily for an 80-acre well. A market demand factor is applied across the board, for example, 30 per cent of the full allowable for all wells. The point here is that such formulas completely ignore M.E.R. capacities and reserves. Two wells at the same depth and with the same spacing will have the same allowables, yet the M.E.R. of one may be several times that of the other. The same is true for reserves. One such well may have many times the reserve of the other but be produced at the same allowed rate.

This difficulty, it must be noted, has not prevented the National Petroleum Council from periodically estimating the producing capacity of the industry by regions and nationally, and estimating a short-run decline factor. It was, indeed, this fact and the further process of assigning reserves back to the years of discovery of reservoirs, which provided Mr. Zapp with the clue to his proposed method of estimation. His point was that, if it can be done once, it can be done every year; and if it is done every year, informative trend values can be established. If the obstacles outlined in the preceding paragraph are taken at full value, questions also arise concerning the validity of the NPC estimates. If the NPC procedures have a substantial degree of validity, then it can hardly be said that Mr. Zapp's proposal is incapable of being put into effect up to a comparable degree of validity, if the industry cared to go to the trouble and expense which would be involved.

Another line of criticism of Mr. Zapp's proposal is that it is applicable only to the estimation of primary reserves, and provides no approach to the estimation of additional reserves which might be obtained through the application of secondary recovery methods. Still another criticism is that, under the dynamics of the industry, changing technology may affect producing capacity without equivalent effect upon reserves.

In conclusion, we may repeat that Mr. Zapp did not attempt to add to quantitative knowledge concerning the volume of reserves, but only to suggest a method by which such knowledge could be improved. Apart from the details of his method, he was undoubtedly right on one point: namely, that improvement of knowledge concerning future availability of oil depends upon data which provide the basis of improved statistical inference in the form of trend analysis.

7

Approaches to the Estimation of "Ultimate Reserves"

The estimating procedures of the API, as we have seen, are earthbound to "existing economic and operating conditions," and even in this context are further restricted by other limiting assumptions. They do not, therefore, open up any vista of the productive potentialities of known fields under less limiting assumptions or of the prospects for future discoveries. For certain purposes this wider view is desired, especially in the case of public agencies with some responsibility for assessing the future energy requirements of the American economy and of considering policies appropriate to changing circumstances.

Out of this interest have arisen a number of studies concerned with the problem of "ultimate reserves." Even for known fields, the obstacles to estimation are serious. The geological properties and extent of the fields are not fully known; the full potentialities of improved rates of recovery are not known; and the future economic factors bearing upon rates of recovery are not known. Beyond these realms of uncertainty, the imagination of some investigators extends to even further reaches of the unknown, seeking some reasonable order of magnitude for the oil in place not yet discovered and for the ultimate rate of recovery.

The term "ultimate reserves" is used in more than one way, which makes it necessary to note the usage of each estimator in order to avoid ambiguity. For our purposes we shall use the term in a loose and inclusive sense to cover all speculation upon the discoverable amount of oil and upon the recoverable portion of it, keeping "ultimate reserves" in quotation marks as a reminder of this loose usage. The elements of the estimating problem, allowing for different treatments, may be stated as follows:

1. The starting point in every case will be (a) the oil heretofore produced and (b) the existing proved reserves in known reservoirs. These are firm figures.

2. To this will be added an estimated amount of reserves to be discovered and/or proved in the future under the limiting assumption of the current state of technology and economic conditions. These are speculative figures.

3. The limiting technological assumption under item 2 above can then be relaxed, and allowance made for additions to reserves arising from improved methods of recovery, both (a) from known reservoirs and (b) from reservoirs yet to be discovered. This stage allows for a range of speculations about rates of recovery.

A number of estimates of "ultimate recovery" have been based upon items 1 and 2 above, and this has perhaps been the most widely used meaning of the term. But any estimate designed to give effect to changing technology must go on to item 3. Varying economic assumptions may be introduced.

THE RESOURCE BASE AND RATE OF RECOVERY: THE RFF STUDY

Over the past ten years there have been a number of studies concerning future possible additions to reserves, each of which implies some original amount of discoverable oil in place and some assumed rate of recovery. These various estimates have been summarized and subjected to critical examination by Bruce C. Netschert in a study sponsored by Resources for the Future, Inc.[1] His analysis offers a convenient starting point for anyone desiring an introduction to this technical literature. Since these speculations on "ultimate" oil lie outside the main purpose of the present paper, we shall not ourselves enter into the various estimates; but we may usefully rely on Mr. Netschert's analysis to establish certain orders of magnitude.

---

[1] For citations on these earlier estimates, see Sam H. Schurr and Bruce C. Netschert, *Energy in the American Economy 1850-1975* (Baltimore: Johns Hopkins Press for Resources for the Future, Inc., 1960). References to studies on various categories of reserves including "ultimate reserves" up to 1959 can be found in Schurr and Netschert, *ibid.*, pp. 350-51. Additional studies since 1959 not noted elsewhere in this study include: Milton F. Searl, *Fossil Fuels in the Future*, U. S. Atomic Energy Commission,

The ultimate reserves to be proved, from the beginning to the end of the domestic oil industry, have, for the most part, been estimated within the range 140 to 250 billion barrels. Since these estimates are generally based upon current estimated rates of recovery, about one-third of oil in place, Mr. Netschert multiplies the reserves estimates by three to get inferred amounts of discoverable original oil in place. These amounts then fall in the range of 420 to 750 billion barrels.[2]

Subtracting from these figures the oil in place already discovered as estimated by Paul D. Torrey, as of January 1, 1960, 328 billion barrels,[3] the oil in place yet to be discovered falls in the range of 92 to 422 billion barrels.

Then, subtracting from the estimates of total discoverable oil in place the amount already produced to 1960, 63 billion barrels, gives a present resource base from which to recover future oil in a range from 357 to 687 billion barrels. Out of these figures Mr. Netschert settles upon 500 billion barrels as a credible resource base of present oil in place, consisting of 265 billion barrels in known deposits and 235 billion barrels yet to be discovered. The 265 billion barrel figure is arrived at by deducting past production (63 billion barrels) from the Torrey estimate of oil originally in place (328 billion barrels). The 235 billion barrel figure is arrived at by deducting 265 from 500.

From this resource base it is possible to compute hypothetical amounts of recoverable oil, by applying any assumed rate of recovery. Mr. Netschert does not specify any expected rate, but he does think that past estimates

---

Office of Technical Information, TIO-8209 (October 1960); Paul Averitt, *Coal Reserves of the United States—A Progress Report, January 1, 1960*, U. S. Geological Survey, Bulletin 1136 (Washington: Govenment Printing Office, 1961); U. S. Atomic Energy Commission, *Civilian Nuclear Power*, Appendices to A Report to the President—1962 (Office of Technical Information, 1962); T. W. Nelson, "Wanted: 100 Billion Barrels of Oil in North America Between Now and 1984," paper presented to the Southwestern District, Division of Production, American Petroleum Institute, Midland, Texas, March 19, 1964; Warren B. Davis, "Is There Enough Oil," paper presented to the Annual Meeting of the A.I.M.E., New York, February 18, 1964; Paul D. Torrey, C. L. Moore, and G. H. Weber, "World Oil Resources," paper presented to the Sixth World Petroleum Congress, Frankfort/Main, June 1963; *Oil and Gas Journal*, "USGS Estimates 4-Trillion-bbl. Reserve," September 16, 1963; reporting on an unpublished study by A. D. Zapp of the Geological Survey; U. S. Department of the Interior, Energy Policy Staff, "Supplies, Costs, and Uses of the Fossil Fuels," February 1963 (mimeo.); Federal Council for Science and Technology, Executive Office of the President, *Research and Development on Natural Resources* (Washington: Government Printing Office, 1963), pp. 43-47.

[2] See Schurr and Netschert, *op. cit.*, p. 358.

[3] The figure of 346 billion barrels which we have used elsewhere is Mr. Torrey's later estimate of January 1, 1962.

of potential recovery have been much too conservative, because they assume existing technology. "Much higher recovery possibilities are indicated by current research and development with new techniques."[4] He suggests what he regards as reasonable orders of magnitude. "A doubling, for example, is wholly plausible. A recovery factor of two-thirds as the average for the United States within the next fifteen years or so would be well within the possibilities currently being opened up."[5] Applied to his resource base of 500 billion barrels, this would yield "reserves" of 156 billion barrels in known fields[6] (including "proved" reserves) and 157 billion barrels in fields yet to be discovered. Similar calculations can be made for any assumed rate of recovery.

Some lines of thought to which Mr. Netschert's analysis give rise may be illustrated by a series of disconnected quotations from another RFF study, *Resources in America's Future.*[7]

> A doubling of historic recovery rates, principally on the basis of greatly expanded secondary activity, would, on the basis of the RFF resource-base judgment, yield some 300-350 billion barrels. A first calculation, therefore, starts with an assumption of two-thirds recovery, or recoverable supplies of 330 billion barrels. To obtain such an increase in recovery as an average over the entire forty years would require nearly 100 per cent recovery toward the final years, but it could also prevail by large initial increases in recovery, followed by stability at a high level short of 100 per cent.
>
> We have, as an alternative, assumed a more limited future availability of 250 billion barrels, equivalent in the next forty years to the recovery of half the oil now estimated to be in place (in proved reserves, unrecovered oil fields, and fields yet to be found), even though we are aware that the estimate may turn out to be a minimum one.
>
> . . . It is quite conceivable, indeed probable, that estimates of the base itself will in the future undergo enlargement. . . . Thus, the 250 billion barrels here assumed . . . might be drawn not at 50 per cent recovery from 500 billion barrels but, say, at 45 per cent from 550 billion barrels, or 40 per cent from 625 billion . . . .

[4] See Schurr and Netschert, *op. cit.*, p. 380.

[5] *Ibid.*, p. 383.

[6] Calculated as follows: 328 billion barrels originally in place times $2/3$ = 219 billions minus 63 billions already produced = 156 billions. This is our calculation, not Mr. Netschert's, and is not meant to suggest that he would expect old fields to be brought up to the same rate of recovery as later fields.

[7] Hans H. Landsberg, Leonard L. Fischman, and Joseph L. Fisher. *Resources in America's Future: Patterns of Requirements and Availablities, 1960-2000* (Baltimore: Johns Hopkins Press for Resources for the Future, Inc., 1963), pp. 390-91.

The magnitude of recoverable supplies presented above does not rest on any level of costs or technology that we can specify. Yet the availability—as opposed to the physical occurrence—of oil today, tomorrow, and forty years hence will depend on the conditions under which the producers will find it profitable to exploit the known and explore the unknown.

Even if we were to assume that advancing technology will successfully overcome the physical obstacles to finding and lifting oil, we would have to determine the time schedule within which the exploitation of existing resources is likely to take place. Since the occurrence of oil, whatever its magnitude, is ultimately finite, exploitation should reach a peak—or perhaps several peaks or an extended plateau—then subside and terminate. . . .

In any attempt to speculate upon the future availability of oil, it is evident how many variables and unknowns there are to deal with—the size of the resource base, technological factors affecting discovery and recovery, economic factors of cost and price, timing of demand, and so on. No sensible investigator has any illusion of providing precise answers. But there is felt to be a necessity to formulate certain orders of magnitude to be compared with projections of demand as a basis for forward thinking about the intensity and imminence of national energy problems.

The procedures and conclusions of this RFF study support expectations concerning future availability of oil wholly divorced from the limitations of proved reserves estimates. Concentration on the proved reserves picture may give the impression of imminent shortage of oil. The "resource base" picture greatly extends the time perspective within which the problems of oil availability can be dealt with. Assuming the orders of magnitude to be credible, how long the oil will last can be computed by estimating future rates of production.

"ULTIMATE RESERVES" AS ESTIMATED BY C. L. MOORE

Mr. Netschert's analysis, reviewed above, does not include any original method of his own for estimating the ultimate availability of oil. Out of the various studies of other experts, he is content to distill what seems to him a reasonable magnitude for the resource-base from which future oil is to be recovered and also reasonable expectations concerning future rates of recovery. By way of contrast, a study by C. L. Moore,[8] prepared in the

---

[8] C. L. Moore, *Method for Evaluating U. S. Crude Oil Resources and Projecting Domestic Crude Oil Availability* (Washington: Department of the Interior, Office of Oil and Gas, May 1962).

U.S. Department of the Interior, devises a mathematical formula for projecting future rates of discovery and recovery. Mr. Moore's method is to establish trend relations by which (1) to estimate the total oil potentially recoverable from presently known fields at calculated rates of recovery, (2) to forecast the amount of oil in place still to be discovered, and (3) combining the other two, to put a figure on recoverable oil from all deposits in the United States, both presently known and as yet undiscovered.

The steps in his analytical procedure are as follows:

1. The first step is to calculate the historic pattern of the discovery of crude oil in place and to project this pattern into the future. The starting point of the calculation is the IOCC (Torrey) estimate of crude oil originally in place discovered up to January 1, 1960 (328 billion barrels). This total is broken up into annual amounts by correlating it basically with the National Petroleum Council (NPC) figures of cumulative proved reserves assigned to the year of discovery (see Table 4 above). Further adjustments are made to the similar data of the Petroleum Administration for War (PAW) 1945 report and for certain other factors, such as the extraordinary discovery peak of 1930. The annual data are reduced to a smoothed curve through which a trend line is drawn. This trend line of annual discoveries of original oil in place is upward to 1930, downward thereafter. Extended by exponential formula to the final zero point, it yields a result of 486 billion barrels of oil originally in place, of which 328 billions have been discovered, leaving 158 billion barrels still to be discovered.

2. The second step is to establish the historic pattern of the per cent recovery of the crude oil originally in place. This percentage at the first of any year is the ratio of cumulative gross additions to API proved reserves to the cumulative discoveries of oil in place as of that time. Starting from a commonly estimated 15 per cent up to 1930, the annual percentages rise to 28.8 per cent in 1960. The calculated trend rises to 43.0 per cent in 1975 and to 57.8 per cent in 1990. An ultimate recovery of 75 per cent is projected.

3. The third step is to synthesize the results of the first two steps to yield data on the cumulative and annual *gross* additions to reserves and to provide a projection of such data. The projections are broken into stages. Gross additions to reserves, 1960–75, are placed at 68 billion barrels, 1975–90 at 76 billion barrels. After 1990, projecting to the ultimate 75 per cent rate of recovery, 125 billion barrels would be added to reserves.

4. The fourth step is to establish the historic pattern of cumulative *net*

additions to reserves (proved reserves as of any date) and to provide a formula for projecting it. This involves establishing the pattern of crude oil production, which can be deducted from the gross additions in the preceding step to yield a net addition to reserves. The calculated net additions to reserves (proved reserves as of a given date) rise to a peak of 44.7 billion barrels in 1983 and then start a slow decline. Compared with actual net addition, 1945–60, of 9 billion barrels, calculated additions are 11 billion barrels for 1960–75 and 1 billion barrels 1975–90. As compared with actual annual rate of withdrawals (production) of 2.5 billion barrels as of January 1, 1960, calculated annual withdrawals are placed at 4.76 billion barrels in 1975 and 4.95 billion barrels in 1990.

The analysis is broken down into estimates of ultimate recovery (1) from presently known fields and (2) from fields yet to be discovered. With respect to (1), the method is simply a mathematical application of the formula of rising rates of recovery applied to Torrey's estimate of original oil in place. The results are staged on the following lines:

Stage 1. Original oil in place in known fields is estimated (by Torrey) at 328.4 billions, as of January 1, 1960. Oil produced had been 62.9 billion barrels and proved reserves were 31.7 billion barrels, for a total of 94.6 billion barrels cumulative gross additions to reserves. This yields an anticipated recovery of 28.8 per cent of oil originally in place.

Stage 2. As of January 1, 1975, the percentage of recovery is projected at 43.0 per cent. Applied to original oil in place (328.4 billion barrels), this yields cumulative gross additions to reserves of 141.2 billion barrels. Subtracting cumulative gross additions to 1/1/1960 (94.6 billion barrels) leaves 46.6 billion barrels to be added to cumulative gross reserves in the fifteen-year interval, 1960-75, as a result of improved recovery in presently known fields.

Stage 3. As of January 1, 1990, the percentage of recovery is projected at 57.8. Applied to original oil in place this yields cumulative gross additions of 189.8 billion barrels. Subtracting cumulative reserves to 1/1/1975 (141.2 billion barrels) leaves 48.6 billion barrels to be added to reserves in known fields in the fifteen-year interval 1975-90.

Stage 4. An ultimate percentage of recovery is projected at 75 per cent. Applied to original oil in place, this yields ultimate cumulative gross additions to reserves of 246.3 billion barrels. Subtracting cumulative reserves to 1/1/1990 (189.8 billion barrels) leaves 56.6 billion barrels to be added to reserves in known fields after that date.

The results may be summarized in this way:

Production to 1/1/1960	62.9 billion barrels
Proved reserves as of 1/1/1960	31.7 billion barrels
Cumulative subtotal to 1/1/1960	94.6 billion barrels
Additions to reserves 1960-75	46.6 billion barrels
Cumulative subtotal to 1/1/1975	141.2 billion barrels
Additions to reserves 1975-90	48.6 billion barrels
Cumulative subtotal to 1/1/1990	189.8 billion barrels
Additions to reserves after 1/1/1990	56.6 billion barrels
Ultimate total	246.4 billion barrels

Turning to oil yet to be discovered, extending his trend line of discovery by exponential formula, Mr. Moore arrives at 486 billion barrels as the amount of discoverable oil originally in place. (This may be compared with Mr. Netschert's present resource base of 500 billion barrels plus 63 billion barrels already produced, or a total of 563 billion barrels.) Deducting 328 billion barrels originally in place in known fields (the IOCC figure), Mr. Moore arrives at 158 billion barrels yet to be discovered. Mr. Netschert's comparable figure is 235 billion barrels.

The recoverable oil from this additional oil in place is staged according to the same formula of increasing rate of recovery as in the case of presently known fields. The composite result is shown in the following tabulation, in billions of barrels.

	1960-75	1975-90	After 1990	Total
From known fields	46.6	48.6	56.6	151.8
From discoveries	21.8	27.5	68.4	117.7
Total	68.4	76.1	125.0	269.5

It is to be understood that these figures give effect in each stage to the rising rate of recovery, from 28.7 per cent in 1960, to 43.0 per cent in 1975, to 57.8 per cent in 1990, to 75 per cent ultimately.

Adding to the above figures the cumulative reserves added before 1960, 94.6 billion barrels (62.9 past production plus 31.7 API proved reserves), the total of "ultimate reserves" comes to 364.1 billion barrels at an assumed recovery rate of 75 per cent.

When staged in this fashion, Mr. Moore's projected values for the early future, to 1975, are fairly in line with various expert opinions within the industry, arrived at by entirely different methods. The then President of the Humble Oil Company, Morgan J. Davis, stated in 1958, for example, that in his opinion at least 70 billion barrels would be added to gross reserves during the next twenty years, suggesting the figures 20 to 25 billions from extensions and revisions in known fields, 20 billions from improved secondary recovery methods, and possibly 35 billions from new discoveries.[9] Some other industry opinions run substantially lower. It is only in the more distant and speculative future, when his rising rates of recovery take hold, that Mr. Moore's figures balloon beyond the possibility of checking with recent experiences.

Mr. Moore carries his projections one stage further to the calculation of future staged withdrawals (production) and the size of proved reserves as of 1975 and 1990. Although production is calculated to rise from 2.5 billion barrels in 1960 to 4.95 billion in 1990, additions to reserves run ahead, leaving net additions to reserves (proved reserves) at that date 12 billion barrels larger than in 1960. Thereafter, a decline sets in.

After making the projections summarized above, Mr. Moore goes on to say: "Before accepting such projections as realistic they should be tested against the operating and economic feasibility of fulfillment." For evidence on the operating side he uses the annual reports of the Committee on Statistics of Exploratory Drilling of the American Association of Petroleum Geologists. Running down the data on number of exploratory holes drilled, the success rates of productive wells (adjusted to natural gas and condensate wells), and the average crude oil in place credited to each productive exploratory well, he projects a decline in average discovery of oil in place per well, which, however, would be offset by an increasing rate of recovery. He concludes that, to support his projections of discovery of oil in place, it would be necessary to step up the drilling of exploratory wells radically, with an annual total perhaps in the range of 15,000 to 20,000 per year, as compared with the peak of 16,173 in 1956 and only 11,704 in 1960.

Starting from the presumption that the funds for additions to reserves are derived mainly from the gross revenue from the sale of oil, Mr. Moore traces, as evidence on the economic side, the course of gross revenue over time in relation to the course of Gross National Product. For the fifteen-year period 1945–60, average annual production of 2,206 billion barrels at an average price of $2.55 per barrel gave average annual gross revenue

[9] Morgan J. Davis, "The Dynamics of Domestic Petroleum Resources," paper presented at the 38th Annual Meeting of the American Petroleum Institute (Chicago: November 12, 1958).

of $5.622 billions. For the period 1960–75, an average annual production of 3.812 billion barrels is projected at an average price of $3.18 to yield average gross revenue of $12.114 billions.

For the period 1945–60, gross revenue grew at an average annual rate of 6.36 per cent, as compared with a rate of 5.86 per cent for GNP. For the period 1960–75, the average annual growth rate of gross revenue is projected at 5.83 per cent, as compared with 4.5 per cent for GNP.

The projected GNP, production, and gross revenue are stated as maximum conditions. If GNP failed to achieve this maximum growth rate, it would be expected that the production and gross revenue would be correspondingly lower.

For 1945–60, gross revenue was 1.669 per cent of GNP. For 1960–75, the projected percentage is 1.733.

On evidence of this sort, Mr. Moore concludes that the projected rates of growth in oil discovery and production are not outside the range of economic feasibility, but rest only on the condition that the industry substantially increase its efficiency both in the discovery of new oil deposits and in increasing the recovery from known deposits. Competing sources of energy could also upset the economics of Moore's calculations.

Mr. Moore's mathematical analysis attempts to bridge the immense gap between the API proved reserves on one side and the shadowy world of "ultimate reserves" on the other. Mr. Moore was himself government co-chairman of the NPC study reviewed above—which, it will be recalled, allocated later extensions and revisions in the API estimates of proved reserves back to the year of initial discovery of fields. His study is the first attempt to utilize those data for purposes of trend analysis. If the prospects of the oil industry are to be intelligently considered at any point in time, trend analysis is one of the essential instruments, combined with expert judgment. Mr. Moore leaves judgment out and does what he can to force the available data into the trend mould. The only basic data he had to work with were the API figures as rearranged in the NPC Report, together with production data and the IOCC (Torrey) estimate of original oil in place in known fields.

One may perhaps be permitted to doubt whether the data are sufficient to support the superstructure of mathematical analysis and statistical projection built upon them. In the first place, the NPC Report had itself disclaimed that its figures would support a proper analysis of discovery trends.[10] Even if they did, it is questionable that such trends could usefully be projected into the distant future. Shorter trend projections developed from relatively recent data are probably the only significant ones.

[10] See *NPC Report, 1961*, pp. 3-4.

Nevertheless, unless one considers the whole realm of speculation on future discoveries and rates of recovery a losing game from the start, Mr. Moore is on the right track in seeking to develop techniques of trend analysis. For that purpose, the immediate need appears to be the development of basic data in a form which will support such analysis.

"ULTIMATE RESERVES" AS ESTIMATED BY M. K. HUBBERT

Another recent attempt to estimate "ultimate reserves" by mathematical analysis from statistical data is that of M. King Hubbert in his recent report, *Energy Resources*.[11] His mathematical-statistical method of estimating the recoverable oil in American deposits is much simpler than Mr. Moore's, consisting for crude oil simply of extrapolations from two statistical series: (1) the cumulative production of crude oil in the United States and (2) the cumulative additions to proved reserves as estimated by the API. (Separate calculations are made for natural gas and natural gas liquids.) The method does not, as in the case of Mr. Moore, permit separate estimates of the anticipated recovery from presently known fields and fields yet to be discovered.

The cumulative gross additions to reserves, called the "cumulative proved discoveries," are charted on a time base. Through the jagged line of actual "rates of proved discovery," is drawn a calculated trend line. This "analytical-derivative curve" reaches a peak in 1956 and then describes a downward course to the right symmetrical to the upward course on the left.

Cumulative production is charted in the same manner, and reaches the size of "cumulative discoveries" with a lag of about ten and one-half years, thus reaching its peak about 1966–67. Proved reserves thus start to decline —as cumulative production exceeds cumulative additions to reserves—at the mid-point, or about 1961–62.

By this method of calculation, cumulative production of crude oil from the beginning to the end of the industry comes to 170 billion barrels. A separate analysis, based on the *Oil and Gas Journal* series on giant and total fields, yields what Mr. Hubbert regards as a corroboratory figure of 175 billion barrels.

Taking 175 billion barrels as his base and deducting the amount produced to the end of 1961 (67.37 billion barrels) and proved reserves (31.76

[11] M. King Hubbert, *Energy Resources: A Report to the Committee on Natural Resources of the National Academy of Sciences-National Research Council*, Publication 1000-D (Washington: National Academy of Sciences-National Research Council, 1962).

billion barrels) leaves him with 75.9 billion barrels of "undiscovered reserves of crude oil." This covers both the unproved content of known fields and the content of undiscovered fields. If, as is commonly supposed in expert quarters, more or less 30 billion barrels will be proved in existing fields under API estimating procedures, and something more added by expanded and improved secondary recovery methods, Mr. Hubbert leaves little oil still to be discovered.

In arguing for the validity of his "cumulative discovery curve," Mr. Hubbert writes as follows:

> [It] is the embodiment of the results of all the improvements which have been made in discovery techniques, in drilling techniques, and all the oil added by geographical extensions within the United States and its offshore areas, since the beginning of the industry. Thus we do not have to worry about how much oil may be contained in known oil fields over and above the API estimates of proved reserves, or how much improvement may be effected in the future in both exploration and production techniques, for these will all be added in the future, as they have been in the past, by revisions and extensions in addition to new discoveries. And there is as yet no evidence of an impending departure in the future from the orderly progression which has characterized the evolution of the petroleum industry during the last hundred years.[12]

By thus staking his whole analysis on the continued validity of the trend exhibited in one catch-all statistical series, cumulative gross additions to reserves, he evades the necessity of looking at any of the separate factors which might be relevant, such as rate of recovery, backdating extensions and revisions to the year of discovery of fields, intensity of discovery effort, and discoveries relative to discovery effort.

Cross-comparing with Mr. Moore, it will be recalled that by a projection of the recovery trend he succeeded in adding 95 billion barrels of hypothetical reserves *in known fields alone* by 1990, or roughly 19 billion barrels more than Mr. Hubbert concedes to the whole future of the industry.

On more technical grounds, we must recall Mr. Zapp's argument (Chapter 6) that gross additions to reserves cannot properly be interpreted as a discovery rate in any sense. Mr. Moore does not do so, but builds up his discovery rate on the backdating of extensions and revisions in the NPC report. Mr. Hubbert proceeds as if gross additions to reserves in the API series can be regarded as expressing a discovery rate, without men-

---

[12] *Ibid.*, p. 60.

tioning the contrary views of Mr. Zapp, Mr. Moore, or the NPC report. The present authors find it impossible to take seriously an estimate of the future availability of crude oil based solely on a mathematical formula for projecting the statistical history of cumulative additions to gross reserves.

More broadly, the experiments of both Mr. Moore and Mr. Hubbert in applying mathematical-statistical techniques to the available data of the petroleum industry raise two questions: (1) whether the data now provided by the industry are adequate as raw materials for achieving useful predictive results from the application of these techniques; and (2) whether, in an industry so beset with uncertainties as the petroleum industry, statistical trends, even if based on superior data, can be usefully extrapolated for more than a relatively short period of years.

On the first point, the present authors consider the available data defective and ill-suited to producing valid projections of trend. On this point, however, a distinction must be made between the two studies. Mr. Moore utilizes all the data available and his study must be regarded as a useful attempt to advance knowledge by employing trend analysis. On the other hand, Mr. Hubbert's work with numbers and techniques appears to add nothing to the embryonic science of "petroleumetrics."

Concern with analytical methodology is necessary in order to have good methods to apply to good data to secure significant results. There is therefore no reason to complain of methodological experiments. It would merely be unfortunate if, looking only at the results, people were to think that those we have reviewed throw any real light on the long-run prospects of the petroleum industry.

On the second point, we side with Mr. Zapp in preferring relatively short-run trend analysis from relatively recent statistical data. We see no purpose that can usefully be served by pursuing the industry to its hypothetical graphical end, even if the supporting statistical data were much superior to what they actually are.

At best, the attempt to make projections by mathematical formulas will face a disconcerting fact. Trend analysis inevitably embodies the economic factor, in particular the changing strength of the incentives to search for domestic oil. Incentives, in turn, are affected by the availability of competitive sources of energy, including foreign sources of oil supply. Hence, the extrapolation of trends yields a hybrid economic-geological estimate of what the earth's crust may yield. This need not prevent the development of useful trend analysis, since the mathematical extrapolation of statistical data can be interpreted in the light of informed judgment concerning possible changes in the economic and technological factors.

ADAPTATIONS OF RESERVES ESTIMATES

In recent years two committees of the United States Congress engaged in examining potential energy sources have set up study groups which have reported, among other things, upon the petroleum resources of the United States. These reports consist partly of the types of information we have reviewed above on potential recovery from presently known fields, partly of kinds of data reviewed under the heading of "ultimate reserves."

The McKinney Report to the Joint Committee on Atomic Energy

In 1960 Robert McKinney headed a committee which prepared a volume of "background material" for the Joint Committee on Atomic Energy.[13] This compilation included some material on the future availability of petroleum, such as a report from the U.S. Department of the Interior on recoverable oil from known fields; the report is based on the IOCC (Torrey) estimates of original oil in place, adjusted to include natural gas liquids and to include Canadian sources. On this expanded basis, past extraction is placed at 80 billion barrels, leaving 300 billion barrels in place in drilled portions of known reservoirs and 75 billion barrels in possible extensions of known reservoirs.

No attempt is made to estimate the eventual recovery from these underground supplies for the following stated reasons. "Just how much will become producible will depend upon technologic progress and economic conditions in the future";[14] and "because the extent of recovery of these oil and gas underground resources is subject to much change resulting from technologic progress and changing economic conditions, there is a growing tendency to give greater emphasis to the total underground quantities in long-range considerations."[15] Minimally, it is suggested that future additions to proved reserves will at least equal present proved reserves, since "The history of proved reserves estimates indicates that extensions and revisions applicable to cumulative discoveries at a given time will more than equal the proved-reserve estimate at that time. Thus

[13] *Background Material for the Review of the International Atomic Policies and Programs of the United States*, report to the Joint Committee on Atomic Energy, Vol. 4, Joint Committee Print, 86 Cong., 2 sess. (Washington: Government Printing Office, 1960).

[14] *Ibid.*, p. 1527.

[15] *Ibid.*, p. 1528

future extensions and revisions for fields now known should at least equal 41 billion barrels of liquid hydrocarbons."[16]

Upon the basis of interviews with members of the industry, a special task force reported that 65 billion barrels of reserves could be regarded as "proved" in the United States, as against the official API figure of 38 billion barrels, for all liquid hydrocarbons, as of the end of 1959.

The Study Group Report to the Senate Committee
on Interior and Insular Affairs

In 1962 a study group set up for the purpose of assessing information on energy in the United States presented its report to the Senate Committee on Interior and Insular Affairs.[17]

The study covers the whole field of energy sources, as to both future requirements and supply. In the relatively brief portion devoted to the future availability of liquid hydrocarbons, the study presents no experiments in the methodology of estimating, being simply an attempt (for the benefit of policy makers) to distill from existing sources some basis for judging the trend of availability of oil in the calculable future.

The sources available to it did not go beyond those already reviewed in earlier chapters of the present paper. The study reached the conclusion that the API estimates of proved reserves were not suitable for its purposes. "The [API] figure has little significance in the present context . . . first because it depends on definition, and second because it conveys little information on how much oil can be made available domestically to meet energy demands"[18]—a view which will be shared by all knowledgeable people.

For a larger figure, it went to the McKinney Report, reviewed above, noting that the McKinney task force "had been advised, in interviews with the industry, that reserves at the end of 1959 amounted to 65 billion barrels, although all this had not yet been 'completely developed'."[19] The study then proceeded to the statement that "there must be some additional quantity [above 65 billion barrels] also, namely, the oil that those others

[16] *Ibid.*, p. 1530.

[17] *Report of the National Fuels and Energy Study Group on An Assessment of Available Information on Energy in the United States,* to the Committee on Interior and Insular Affairs, U.S. Senate. Committee Print, 87 Cong., 2 sess. (Washington: Government Printing Office, 1962).

[18] *Ibid.*, p. 52.

[19] *Ibid.*, p. 64.

felt to be too uncertain to be classed as proved even by them . . . we know of no estimate of how much that may be."[20]

As a basis for speculating upon the size of this larger unknown quantity, the study fell back upon the IOCC (Torrey) studies reviewed above, (citing the report as of January 1, 1960 rather than that of 1962—not then available). Rounding the IOCC figures of crude oil currently in place to 250 billion barrels, and adding 20 billion barrels for natural gas liquids, the study speculated upon what recovery may be obtained from this total of 270 billion barrels in place. Recovery of at least 70 billion barrels is deemed to be assured. The question then is how much of the remaining 200 billion barrels can be recovered. The answer depends upon technological and economic factors, the report merely noting that, if recovery could be increased from one-third to two-thirds of estimated oil originally in place, more or less 100 billion barrels of additional oil would be forthcoming.

As to "ultimate reserves," the staff study made a calculation, based on a Geological Survey study, which yielded a speculative result of 270 to 370 billion barrels of recoverable oil in as yet undiscovered fields.

[20] *Ibid.*, p. 65.

II

*METHODS OF ESTIMATING
RESERVES OF NATURAL GAS
AND NATURAL GAS LIQUIDS*

8

Estimating Procedures for Gas Reserves

Much of what was said in Part I about the API estimating procedures for crude oil reserves applies to the natural gas reserves data collected by the American Gas Association (AGA). Gas reserves figures are reported by the AGA in the same general way that crude oil reserves are reported and appear annually in a document issued jointly by the two agencies.

The AGA Committee on reserves was set up immediately after World War II (1945) when it became apparent that the technological and economic barriers to long-distance gas transmission had been overcome and that the large concentration of population and industry in the midwestern and eastern portions of the nation could be supplied economically with gas from the southwestern producing regions. The impetus for gas reserves data must have come in part from the financial community, which was being asked to lend hundreds of millions of dollars to newly formed or expanding gas transmission companies. Also, Federal Power Commission (FPC) pipeline certification requirements include, among other things, that there be evidence that an adequate supply of gas reserves is dedicated to a proposed line.

COMMITTEE PROCEDURES AND MEMBERSHIP

The procedures of the AGA Committee and its subcommittees are described in the annual report on reserves only in a very general fashion. The industry literature does not include a full description of these activities. A brief description, quoted at length below, was given by C. E. Turner, manager of the economics department of the Phillips Petroleum Company, in testimony before the Federal Power Commission.

The board of directors of the American Gas Association, from time to time, appoints a chairman of the Gas Reserves Committee from the Industry. Since the inception of the Committee that chairman has been a reserves expert who is a representative of a gas pipeline company. The chairman appointed by the board of directors is authorized to appoint the members of the main committee. Exclusive of the secretary, there are twelve main committee members, each representing a different producing area of the United States. Each of these members serves as chairman of the subcommittee for the producing area that he represents and appoints the members of his subcommittee. There are eight members on my subcommittee for the producing area which includes the Permian Basin Area, as defined by the Order of the Commission setting this hearing. All of the members of the main committee and of the subcommittees are specialists in the matter of estimating reserves of natural gas and natural gas liquids.

The API has a similar main reserves committee and various subcommittees which make estimates of proved reserves of crude oil and its contained dissolved gas and natural gas liquids. In making their annual estimates, the committees and subcommittees of the two organizations work together to arrive at the best possible estimates of proved total reserves of natural gas and natural gas liquids.

Each member of the AGA subcommittees is assigned a portion of the producing area covered by his subcommittee and he reports discoveries, production and reserves for that subarea. Several times during the year and shortly after the end of the reporting year the whole subcommittee meets and reviews the material developed by its various members and finally arrives at the best possible estimate. Its report is then submitted to the main committee.

The American Gas Association has another committee called the Committee on Underground Storage, which is made up of members of the Industry who are experts on gas reserve and underground storage matters. That committee makes its separate study of underground storage and furnishes its reports to the chairmen of the various subcommittees of the main Gas Reserves Committee.

Each year the main committee of the AGA meets with the main committee of the API to consolidate the reserves and production figures and prepare a final report which is published jointly in pamphlet form.[1]

[1] Prepared Testimony of C. E. Turner in the Permian Basin Area Rate Hearing, before the Federal Power Commission, Docket Nos. 61-1, *et al.*, pp. 6-7.

It is clear that co-operation between the API and AGA parent committees and subcommittees is necessary. The makeup of the various committees assists in this liaison. In 1964 there were 143 members of API subcommittees and 94 members and alternates of AGA subcommittees. Twenty-five members during that year served on both gas and oil subcommittees. Undoubtedly, there is far more continuity than is indicated by this overlap since membership among company geologists and engineers rotates irregularly and many company people work on reserves estimates which are presented by a single subcommittee member. In addition, there are joint meetings between AGA and API subcommittees on the exchange of working papers. One desirable aspect of the extensive subcommittee overlap is to provide some degree of consistency in definitions, concepts, and estimating procedures for both oil and gas, to the extent that such similarities are useful. The smaller number of members on the AGA subcommittees reflects, in part, the greater concentration of gas reserves geographically, and probably also the fact that gas reserves estimations are somewhat easier to make.

The composition of the thirteen-man parent AGA Committee, as of 1964, shows an interesting balance between gas transmission and gas producing companies. The membership of the AGA Committee is listed below.

Ed Parkes *Chairman*, United Gas Corporation
W. F. Burke, Lone Star Gas Company
D. S. Colby, U. S. Bureau of Mines
P. A. Cole, Cities Service Gas Company
R. O. Garrett, Texas Gas Exploration Corporation
B. B. Gibbs, Union Producing Company
J. V. Goodman, Equitable Gas Company
J. N. Newmeyer, Gulf Oil Corporation
E. D. Pressler, Humble Oil & Refining Company
D. R. Scherer, Southern Natural Gas Company
M. T. Whitaker, Socony Mobil Oil Company, Inc.
A. H. Wieder, Shell Oil Company
T. I. Gradin, *Secretary*, American Gas Association

Several of the companies represented are primarily producing companies with few, if any, corporate ties to gas transmission companies. Several are primarily gas transmission companies with some, but not extensive, gas production.

It seems clear that large company representatives who have a backup of extensive company work contribute greatly to gas subcommittee work. Mr. Turner, in the FPC testimony noted earlier, points out that Phillips, because of its extensive interests in the Permian Basin, has detailed reserves estimates for this large area, for all properties and not just those owned by Phillips.[2] It is probably also true that there is a greater concentration of ownership of gas reserves, especially nonassociated gas reserves, among the large oil companies than is true in the case of crude oil. This is explained in part by the fact that there is usually very wide well-spacing in gas field development, thus requiring the control of extensive acreage. Also, small oil producers often find it more desirable to sell gas reserves rather than risk the delay in obtaining pipeline connections or FPC approval of interstate sales. In addition to outright ownership of gas reserves, many large oil companies gather raw gas in the field for processing in natural gasoline plants. Companies which do this are apt to have extensive knowledge of the reserves of producers from whom they purchase.

Fifty-two of the ninety-four members and alternates of the 1964 sub-committees of the AGA were from major integrated oil companies. The remaining forty-two were employees of smaller oil companies and natural gas transmission companies or the producing subsidiaries of transmission companies. Transmission companies have extensive information on gas reserves because of their need to know how much gas is dedicated to their specific pipelines. A transmission company will usually make its own estimates of reserves under a gas sales contract before signing the contract with the oil or gas producing company selling the gas.

THE QUANTITATIVE ESTIMATES

The estimates of proved natural gas reserves for the nation as of the end of 1964 are shown in Table 9. The reporting of these data differs in some details from the reporting of crude oil figures, but the over-all result is substantially the same. A discussion of the various categories of reserves shown in the table will follow.

Table 10 shows a time series of gas reserves data from 1945, the date of inception of the AGA Committee, to the present.

Table 11 indicates the state breakdown of gas reserves for 1963. This gives an indication of the relative importance of separate states in gas reserves and production. The top ranking gas reserves states coincide generally with the top ranking oil reserves states, although the particular rank of an individual state may differ as between oil and gas.

[2] *Ibid.*, pp. 9ff.

TABLE 9. NATURAL GAS RESERVES, 1964

(*Thousands of cubic feet at 14.73 psia*[a] *and at 60° F*)

Total proved reserves as of December 31, 1963.................		276,151,233,000
Extensions and revisions of previous estimate during the year of 1964.......................	13,342,837,000	
New reserves discovered in 1964................	6,909,301,000	
Net changes in underground storage during 1964..	195,111,000	
Total proved reserves added and net changes in underground storage during 1964...		20,447,249,000
Total proved reserves as of December 31, 1963 and additions during 1964..		296,598,482,000
Deduct production during 1964.............................		15,347,028,000
Total proved reserves of natural gas as of December 31, 1964......		281,251,454,000

[a] Pounds per square inch absolute.

SOURCE: American Petroleum Institute and American Gas Association, *Proved Reserves of Crude Oil, Natural Gas Liquids and Natural Gas*, December 31, 1964, Vol. No. 19 (New York: 1965), p. 19.

The natural gas reserves figures "*exclude* gas loss due to natural gas liquids recovery." Thus, those parts of the raw gas streams that are recoverable as natural gas liquids (NGL) are deducted from the gas reserves estimates and included in reserves estimates for NGL. Some of the problems in making these adjustments will be taken up in the discussion of NGL reserves.

THE DEFINITION OF PROVED RESERVES

The AGA in its annual report defines its concept of proved reserves in terms which generally appear to follow the definition for crude oil reserves stated by the API Committee. There are, however, one or two major differences which are discussed later.

The definition is as follows:

The Committee wishes to point out that it is often not possible to estimate the total reserves of a field in the year of its discovery. Satisfactory estimates can be made only after there has been sufficient drilling in the fields and, in some cases, adequate production history established. For these reasons, the reserves listed as discovered during any current year must be considered only as the reserves indicated by

TABLE 10. SUMMARY OF ANNUAL ESTIMATES OF NATURAL GAS RESERVES FOR PERIOD DECEMBER 31, 1945, TO DECEMBER 31, 1964

(Millions of cubic feet at 14.73 psia[a] and at 60° F)

Year	Natural Gas Added during Year			Net change in underground storage	Net production during year	Estimated proved reserves as of end of year	Increase over previous year
	Extensions and revisions	Discoveries of new fields and new pools in old fields	Total of discoveries, revisions and extensions				
1945						146,986,723	
1946	b	b	17,632,864	b	4,915,774	159,703,813	12,717,090
1947	7,529,538	3,391,649	10,921,187	b	5,599,235	165,025,765	5,321,952
1948	9,716,426	4,106,664	13,823,090	51,202	5,975,001	172,925,056	7,899,291
1949	8,017,646	4,587,818	12,605,464	82,297	6,211,124	179,401,693	6,476,637
1950	9,122,566	2,861,724	11,984,290	54,006	6,855,244	184,584,745	5,183,052
1951	12,942,930	3,022,878	15,965,808	132,030	7,923,673	192,758,910	8,174,165
1952	8,885,946	5,381,656	14,267,602	197,770	8,592,716	198,631,566	5,872,656
1953	13,298,736	7,043,200	20,341,936	513,626 c	9,188,365	210,298,763	11,667,197
1954	4,607,151	4,939,919	9,547,070	90,412	9,375,314	210,560,931	262,168
1955	16,209,610	5,688,009	21,897,619	87,161	10,063,167	222,482,544	11,921,613
1956	19,110,250	5,605,864	24,716,114	133,242	10,848,685	236,483,215	14,000,671
1957	11,057,936	8,950,119	20,008,055	178,757	11,439,890	245,230,137	8,746,922
1958	13,316,094	5,580,624	18,896,718	57,588	11,422,651	252,761,792	7,531,655
1959	14,852,007	5,769,245	20,621,252	160,450	12,373,063	261,170,431	8,408,639
1960	7,293,016	6,600,963	13,893,979	281,272	13,019,356	262,326,326	1,155,895
1961	10,258,693	6,907,729	17,166,422	159,543	13,378,649	266,273,642	3,947,316
1962	13,184,795	6,299,164	19,483,959	159,230	13,637,973	272,278,858	6,005,216
1963	12,586,733	5,577,934	18,164,667	253,733	14,546,025	276,151,233	3,872,375
1964	13,342,837	6,909,301	20,252,138	195,111	15,347,028	281,251,454	5,100,221

a Pounds per square inch absolute.
b Not estimated.
c All native gas in storage reservoirs formerly classified as a natural gas reserve is included in this figure.
SOURCE: Same as Table 1.

the drilling in that year. The reserves of all fields and pools are reviewed and revised upward or downward in each succeeding annual report to reflect additional information on preceding estimates. These changes are shown as "Extensions and Revisions."

The procedure followed in estimating and assembling the proved reserves figures is the same as that used in past reports. A proved reserve may be in either the drilled or undrilled portion of a given field. When the undrilled area is considered proved, it is so related to the developed acreage and the known field geology and structure that its productive ability is considered assured. Proved recoverable reserves of natural gas are those reserves estimated to be producible under present operating practice, with no consideration being given to their ultimate use. Since the estimates are made by pools, the recovery factor or abandonment pressures used in the calculations are governed by the operating conditions in each individual pool.

Gas reserves data are reported by type of reservoir, that is, nonassociated gas, associated gas, and dissolved gas (See Table 11), and the definition of proved gas reserves must also include the definitions of these three categories. These sub-definitions are not found in the text of the report, but rather in the footnotes to the tables. We shall return to questions raised in the central definition after a digression on classification by type of reservoir.

While the classification of gas reserves in the nonassociated, associated, and dissolved categories conveys a general impression as to the relationship of gas to oil in reservoirs, the task of putting a specific reservoir in a specific category often requires the judgment of expert personnel who are familiar with the technical aspects of reservoir mechanics and behavior. Every reservoir is unique, and the characteristics of each must be examined in detail in the classification process. In this sense gas reserves estimation by category presents problems not encountered in crude oil reserves estimation.

Gas may be found alone in a reservoir and wholly in the gaseous state. Such gas would be classified as "nonassociated." Oil may be found in a reservoir with no gas present except that which is in solution with the oil. This gas would be classified as "dissolved." Gas and oil may be found in a reservoir in many different combinations when the field is discovered, and the relationship of gas to oil may or may not change, depending on reservoir and fluid characteristics and on drilling, completion, and production practices. For example, gas may be found to be in part separate from the oil and in part in solution with the oil. The former may be termed

TABLE 11. Estimated Proved Recoverable Reserves of Natural Gas in the United States

(Millions of cubic feet at 14.73 psia and at 60° F)

	Reserves as of December 31, 1963 b (1)	Changes in Reserves during 1964					Reserves b as of December 31, 1964			
		Extensions and revisions b (2)	Discoveries of new fields and new pools in old fields b (3)	Net change in underground storage c (4)	Net production d (5)	Total (columns 7+8+9+10 also columns 1+2+3+4 less column 5) (6)	Non-associated e (7)	Associated f (8)	Dissolved g (9)	Underground storage h (10)
Alaska	1,690,724	146,058	1,000	0	6,417	1,831,365	1,748,527	0	82,838	0
Arkansas	1,792,644	286,586	97,780	7,165	84,083	2,100,092	1,569,426	317,722	189,191	23,753
California i	8,865,726	672,499	141,808	11,598	637,924	9,053,707	3,354,184	1,647,953	3,849,350	202,220
Colorado	1,876,057	(−) 64,438	18,050	1,313	101,724	1,729,258	1,399,949	79,323	244,421	5,565
Illinois	168,595	(−) 128	80	17,301	6,347	179,501	72		35,047	144,382
Indiana	60,180	2,264	46	12,048	3,270	71,268	607	597	16,883	53,181
Kansas	17,994,235	42,157	57,134	2,579	817,963	17,278,142	16,578,324	428,220	178,777	92,821
Kentucky	1,085,236	5,217	68,776	9,578	74,452	1,094,355	982,153	0	66,978	45,224
Louisiana i	75,364,992	4,915,365	2,950,060	121	4,154,229	79,076,309	65,012,415	9,175,211	4,887,971	712
Michigan	722,812	44,293	14,309	23,566	30,704	774,276	151,853	73,709	53,663	495,051
Mississippi	2,481,627	58,306	10,250	581	195,000	2,355,764	1,907,408	179,547	262,256	6,553
Montana	598,131	6,660	2,176	11,397	28,095	590,269	384,278	21,122	78,165	106,704
Nebraska	100,042	(−) 534	2,771	3,035	11,931	93,383	56,484	8,169	15,352	13,378
New Mexico	15,037,822	1,116,604	85,800	598	886,389	15,354,435	11,463,976	2,114,892	1,749,048	26,519
New York	132,285	(−) 1,147	1,360	3,785	2,794	133,489	38,288	0	31	95,170
North Dakota	1,119,575	19,420	7,884		36,234	1,110,645	6,372	328,340	775,933	0
Ohio	748,187	(−) 22,378	6,000	15,325	37,713	709,421	228,818	0	89,278	391,325
Oklahoma	19,138,820	1,586,206	239,688	149	1,207,628	19,757,235	14,963,987	2,700,233	1,959,732	133,283
Pennsylvania	1,214,498	69,555	14,560	30,284	85,322	1,243,575	703,957	0	18,104	521,514

Texas [i]	117,809,376	4,403,407	3,074,409	4,112	6,436,249	118,855,055	80,457,781	25,010,649	13,309,782	76,843
Utah	1,638,324	(−) 65,955	12,058	64	65,088	1,519,403	882,146	382,106	254,299	852
Virginia	31,303	0	2,800	0	1,923	32,180	32,180	0	0	0
West Virginia	2,311,164	175,859	27,353	30,022	196,453	2,347,945	1,936,618	0	61,202	350,125
Wyoming	3,988,546	(−) 53,775	70,062	(−) 880	235,393	3,768,560	3,223,793	151,545	372,687	20,535
Miscellaneous [a]	180,332	736	3,087	11,370	3,703	191,822	38,764	0	19,005	134,053
TOTAL	276,151,233	13,342,837	6,909,301	195,111	15,347,028	281,251,454	207,122,360	42,619,338	28,569,993	2,939,763

[a] Includes Alabama, Arizona, Florida, Iowa, Maryland, Missouri, Tennessee, and Washington.

[b] Excludes gas loss due to natural gas liquids recovery.

[c] The net difference between gas stored in and gas withdrawn from underground storage reservoirs, inclusive of adjustments and native gas transferred from other reserve categories.

[d] Net production equals gross withdrawals less gas injected into producing reservoirs. Changes in underground storage and gas loss due to natural gas liquids recovery are excluded. Fourth quarter production estimated in some instances.

[e] Nonassociated gas is free gas not in contact with crude oil in the reservoir; and free gas in contact with oil where the production of such gas is not significantly affected by the production of crude oil.

[f] Associated gas is free gas in contact with crude oil in the reservoir where the production of such gas is significantly affected by the production of crude oil.

[g] Dissolved gas is gas in solution with crude oil in the reservoirs.

[h] Gas held in underground reservoirs (including native and net injected gas) for storage purposes.

[i] Includes off-shore reserves.

SOURCE: Same as Table 1.

a natural "gas cap" and be classified as "associated" gas, while the latter would be "dissolved" gas. Under some reservoir conditions and producing practices the dissolved gas may come out of solution and form a secondary gas cap or add to a "natural" gas cap. In the case of an oil reservoir with a gas cap, there may be oil wells producing only a minimum quantity of gas per barrel, and there may exist in the same reservoir gas wells producing gas with no oil. In oil reservoirs with large gas caps the oil will usually be produced from wells completed in the lower portion of the reservoir where the oil is located. Gas wells producing from the gas cap will often have their production severely curtailed until most of the recoverable oil is captured, in order to maintain reservoir pressure. At this point in time, the gas cap wells will have their restrictions lifted and will be produced essentially as a nonassociated reservoir. In some fields, gas may be injected for "pressure maintenance" purposes. Such a procedure may slow the natural pressure decline as the reservoir is produced and may keep dissolved gas in solution to maintain desirable fluid characteristics of the oil being produced.

Two other terms used in the literature, "oil-well gas" and "gas-well gas," do not fit together too well with the above categories. Oil-well gas includes any gas produced along with the oil and is normally dissolved gas, although it could conceivably include some associated gas if a well were producing on the edge of the oil-gas contact. Gas-well gas may be either from a nonassociated gas reservoir or from an associated gas reservoir. There is no way to distinguish between the gas from these two types of reservoirs when it is reported as gas-well gas.

Nonassociated Gas Reserves

Nonassociated gas is defined in the AGA report as gas which:

> is free gas not in contact with crude oil in the reservoir; and free gas in contact with oil where the production of such gas is not significantly affected by the production of crude oil.

Such gas may be "wet," that is, it contains NGL, or it may be "dry," having virtually no liquid content. In either case, oil is present only in small quantities, if at all. Where oil is produced, the gas-oil ratio is so high that gas and natural gas liquids are considered the primary products and oil the secondary product. The occurrence of natural gas liquids in non-associated gas reservoirs is quite common, as shown later in the reserves data for these liquids in Table 11.

Table 11 indicates that roughly three-quarters of the nation's gas reserves fall into the nonassociated category. This proportion has varied only slightly during the postwar years. Since this is the largest part of total gas reserves, its estimation has the most serious implications for policy decisions.

The AGA goes no further in defining this category than the brief wording quoted above. Individuals familiar with reservoir mechanics can perhaps interpret this phrase with the meaning that is intended by the Committee, since they are acquainted with industry practices and concepts relevant to gas-oil ratios and production and with state conservation regulations which deal with this point. Undoubtedly the subcommittees have some more detailed guide lines to determine when "the production of gas is not *significantly* affected by the production of crude oil" (emphasis supplied), but the reader of the report who is not familiar with reservoir behavior is given no indication of what these guide lines are. The borderline between "significantly" and "not significantly" affected by oil production may be vague in the minds of many people who use reserves data, and some further elaboration of the definition would be helpful.

Associated Gas Reserves

Associated gas is defined in the AGA report as gas which:

is free gas in contact with crude oil in the reservoir where the production of such gas is significantly affected by the production of crude oil.

This type of gas is commonly called "gas cap" gas because of the occurrence of a large bubble or cap of gas which, in effect, "floats" on top of the oil in a gas-oil reservoir. The ratios of gas to oil will vary greatly from one gas-cap reservoir to another, depending on the reservoir itself and how it is drilled and produced. In general, good conservation practice usually dictates that associated gas production be curtailed until most of the oil has been captured. As in the case of nonassociated reserves, the rather cryptic definition of associated gas reserves leaves the person who is unfamiliar with the technicalities of reservoir behavior and mechanics without any concept of the specific nature of this reserve. Associated gas reserves accounted for about 15 per cent of total reserves in 1964, as shown in Table 11.

Dissolved Gas Reserves

Dissolved gas is defined in the AGA report as gas which:

is in solution with crude oil in the reservoirs.

Gas is soluble in oil, and in all oil reservoirs the temperatures and pressures are such that some gas is present in solution. The quantities of dissolved or solution gas present may vary greatly among reservoirs. For the dissolved gas reserves estimates the AGA subcommittees rely upon information supplied them by the API subcommittees on the amounts of gas in solution in the various crude oil reservoirs and the producing gas-oil ratios.

Dissolved gas reserves accounted for about 10 per cent of total gas reserves in 1964. (See Table 11.) The remaining small fraction of total reserves is gas in underground storage. This gas is usually brought from a gas producing region to a point fairly close to the point of consumption. Here it is put into an underground storage reservoir, either a depleted oil or gas reservoir, an aquifer, or a limestone or other cavern, and used during periods of peak consumption.

The reporting of gas reserves by type of reservoir or occurrence is helpful because of the unique problems found in each category. The need for expanded information by reservoir type will be brought out later in the discussion.

Returning now to the central concept of proved reserves, we note that a comparison of this definition with that for oil set forth earlier in this study taken from the API report shows several distinctive features. The AGA definition is considerably shorter and somewhat less detailed than that for oil. There are no statements about what is *not* included under the definition, nor does one find the reservations *"beyond reasonable doubt"* and *"under existing economic . . . conditions."*

Since, as we shall see, the concept of proved gas reserves does in fact differ significantly from that of proved oil reserves and that of proved NGL reserves, the discussion of all three in a single pamphlet under the heading of "proved" without cross-comparison tends to be misleading to all but the small group who are intimately acquainted with the estimating procedures used in all three categories. The discussion that follows will attempt to point out significant differences in definitions and note some problems created by these differences.

As in the case of the oil reserves definition, there are some key words in the definition of proved gas reserves which merit comment. For gas the critical words are (1) "reserves estimated to be producible under present operating practices," and (2) "with no consideration being given to their ultimate use."

■ "Reserves estimated to be producible under present operating practices" refers in part, no doubt, to the technology of gas-well drilling and gas recovery. Unfortunately, there is no further explanation of what it means. Comments from industry representatives indicate uncertainty within the industry itself as to what precisely falls within the definition of proved gas reserves. However, one thing seems to be clear. The general view seems to be that the AGA Committee does not exclude a reserve because it is "noneconomic." Gas which is producible under present operating practices in a technical or engineering sense is included. If this is broadly interpreted, it means that gas reserves are reported as proved even though they are not currently being produced because they are noneconomic. Included in this noneconomic category would be such things as gas in remote places such as Alaska, gas in small deposits for which a pipeline is not feasible, and associated gas in a depleted oil reservoir which may be at low pressures and in small quantities. We have no idea what quantities of gas would fall into the producible but noneconomic category. A factor that may keep the amount fairly small is that a gas discovery which is noneconomic will likely not be developed, and since reserves added by an initial discovery are often small, the larger amounts that might be proved via the revisions and extensions will never be reported, or at least not until they become economic. This is a rather anomalous situation, since part of the noneconomic reserves are put into the proved reserves figures and part are left out.

It is possible to conceive of other perplexing situations. Gas in an oil reservoir that is being pressure-maintained may remain in the reservoir until the economically recoverable oil is depleted. The field may then be abandoned with the gas still in place. Presumably this gas is technically producible and is carried as a gas reserve.

It is clearly impossible for the AGA Committee to keep economic criteria from creeping into the proved reserves definition. Present operating practices, even in a narrow technical sense, are partially influenced by existing economic conditions. Gas wells are abandoned before all the gas is recovered, because it costs too much to get out more, relative to the

revenue that could be expected from the incremental gas. At abandonment, presumably, the remaining gas is not included in proved reserves
figures. Yet it is just as logical to include this economically nonrecoverable
gas in old fields in reserves data as it is to include new fields in inaccessible
places. In this hypothetical near-abandonment field, it is quite possible
that a price twice as high as the current price would result in more gas
being recovered from that pool which presumably would be recognized in
reserves data, even given "present operating practices." In addition to this,
present operating practices may change *because* the price incentive has
increased. Secondary recovery of gas is uncommon because primary
recovery usually yields a high percentage of the gas originally in place.
There are, however, situations in which new operating practices have made
what were considered to be noncommercial deposits commercial.[3] The
incentives are clearly economic.

In a sense we may be building straw men in the preceding discussion.
We do not know how much of the current proved reserves of gas could be
considered noneconomic. We suspect that it is not large, but we are not
convinced it is insignificant. Perhaps it is no more than 5 per cent of total
reserves; however this amount is almost equal to a year's production. A
more important fault is the sloppiness of the conceptual apparatus. It is
so nebulous that it means quite different things to different people, and
these people include competent technical people within the industry. It is
certainly true that the difficulties encountered in providing a usable and
understandable conceptual framework are considerable, and may, in fact,
be more difficult for gas than for oil.

It is in a way curious that any reference to "economic conditions" is
avoided in the definition of gas reserves, when economic conditions are
referred to in the oil reserves concept. Discussions with industry representatives yielded no completely satisfactory explanation, but hints at
reasons for the omission were given. Since about one-fourth of the gas
reserves occur jointly with oil and considerably more occurs jointly with
NGL, it is a very complicated problem to talk about the economic conditions relevant to gas alone and not to oil or NGL or to the joint product.
For example, most casinghead gas is recovered today and is either sold or
put back into the reservoir for pressure maintenance purposes. Would this
gas be recovered in the absence of a market for the oil? In many cases it
probably would not, since oil must be produced to get the gas, and since

---

[3] An example of this is the use of formation fracturing of the Picture Cliff and Mesa
Verde sands in the San Juan Basin and the Mancos sands in western Colorado which
has created commercial fields in these areas.

quantities are often relatively small and pressures relatively low. Thus, these gas reserves might not be considered producible and might not be included in the proved gas reserves figures. Therefore, gas may be producible and even economical to recover as a joint product of an oil and gas producing well but nonproducible and uneconomical to recover when not considered a joint product.

To complicate this example even further, we have state conservation agencies which are charged with the duty of preventing "physical waste." Under this mandate, some regulatory bodies have insisted that markets be found for casinghead or oil-well gas before oil production is allowed. There are cases on record in which casinghead gas was given away by a producing company in order to obtain permission to produce oil. Clearly, the "existing economic conditions" surrounding such situations are difficult to sort out. Much gas is "producible" because it is produced with oil. Because it is produced and will continue to be produced as long as the oil is economically recoverable, the gas reserves behind this production are included in "proved recoverable gas reserves." If the market for oil suddenly vanished, there is little doubt that substantial quantities of gas reserves occurring with oil would also vanish. These problems which arise because of the joint product nature of oil and gas are not easily resolved. In the reporting of oil, gas and NGL reserves data they are not even recognized.

■ "With no consideration being given to their ultimate use," the other key phrase in the definition of proved gas reserves is not elaborated in the report. The wording of this phrase is somewhat mysterious, since it can be construed in at least two ways. The phrase could be read as "with no consideration being given to *whether the gas reserves will be used or not used.*" Or it could be read as "with no consideration being given to the type of *end-use to which the gas will be put.*" Apparently the first interpretation is the one intended by the AGA Committee, and not the second. These two interpretations are similar in some respects, but the second seems to make the assumption that only gas reserves which have a use are included in the definition. If we accept the first interpretation as correct, we seem to be back to economic criteria, or the lack of them, in the definition. Any gas reserve which is technically producible is included regardless of whether it will be produced or not and regardless of whether it will be used or not if it is produced. It seems that the AGA Committee has adopted this concept to avoid the insuperable problem of trying to determine the disposition of produced gas, that is, whether it is consumed, reinjected, or flared, or vented. In any case, the concept should be clarified.

There are today small quantities of produced gas which are not used. In some states gas is flared or vented in the field, although state conservation regulations place severe restrictions on this. For example, in 1960 Texas flared or vented in the field about $6\frac{1}{2}$ per cent of the produced oil-well gas. In addition, some gas is "wasted" in cycling and processing plants.

The gas moving to cycling plants and processing plants—natural gasoline and liquefied petroleum gases (LPG) plants—may be oil-well gas, gas-well gas, or a mixture of these. In these plants the liquids are removed from the raw gas stream and sold as final products or as feed stocks to petroleum refineries or chemical plants. Most of the residue gas from cycling plants is injected back into the producing formation to supplement the natural reservoir energies and increase the ultimate recovery of liquids, including oil. In processing plants the liquids are removed and sold, and most of the residue gas is sold to gas transmission lines. Both cycling and processing plants flare or vent some gas, and thus, again, here is a situation of no ultimate economic use for this gas. The commingling of gas streams and the number of steps the gas goes through make it impossible, in many cases, to determine the wells or pools from which the gas comes.

In its proved reserves figures, the AGA reports the reserves from which the "waste" gas, both in the field and at plants, is produced. It might be argued that proved gas reserves should somehow be deflated slightly to account for producible reserves which will be produced and then flared or vented, but just how to do this is not clear. A rough estimate might be gotten by taking "vented and wasted gas" for the nation, reported by the Bureau of Mines, as a percentage of total gas-well and oil-well gas withdrawals minus gas used in repressuring. In 1961, this amounted to about 3.8 per cent. Applied to the total proved reserves figure for 1963 of about 276.2 trillion cubic feet of gas, an amount equal to 3.8 per cent "noneconomic" reserves from a use standpoint would be about 10.5 trillion cubic feet of gas. It should be emphasized that such a calculation provides only rough orders of magnitude. Much of this "waste" would cease if costs of salvaging it were reduced or prices for gas were raised. As recently as 1950, gas "vented and wasted" as a percentage of gross withdrawals was over 9 per cent, excluding "waste" by processing plants. This indicates that much is being done to use the by-product gas which formerly was flared or vented.

The foregoing discussion has explored some *possible* reasons for economic considerations being omitted in the definition of proved gas reserves by the AGA Committee on reserves. To some extent this is speculation, since the AGA is completely silent on this point.

It should be stressed again that the omission of economic criteria for gas reserves, both as to what is economically producible and what is economically usable, creates, in effect, a different concept of proved reserves for gas than for oil. We hasten to add that the economic criteria for crude oil reserves are not at all clear, nor do we know precisely how they are used. It should also be emphasized that applying economic criteria in the case of gas reserves appears to be more difficult than for oil, although not impossible.

CATEGORIES OF CHANGES IN GAS RESERVES

As in the case of crude oil, additions to gas reserves reported annually by the AGA are broken down into reserves coming from (1) "discoveries of new fields and new pools in old fields," and (2) "extensions and revisions." Gas reserves totals for the nation are reported in these two categories only, while oil reserves are reported separately in three categories—discoveries, extensions, and revisions. The definition of proved reserves quoted earlier in this chapter has very little to say by way of explanation of these reserves categories.

■ "Discoveries" are treated in the following statement from the AGA report:

The Committee wishes to point out that it is often not possible to estimate the total reserves of a field in the year of its discovery. Satisfactory estimates can be made only after there has been sufficient drilling in the fields, and, in some cases, adequate production history established. For these reasons, the reserves listed as discovered during any current year must be considered only as reserves indicated by the drilling in that year.

By definition, therefore, gas reserves attributable to discoveries are severely restricted, as in the case of crude oil. A discovery gas well is assigned a certain producing acreage—usually 160, 320, or 640 acres—and the reserves beneath this acreage are assigned as discovery reserves in the year the well is completed. It should be noted that the acreage assigned to a discovery gas well is considerably larger than the acreage assigned to a discovery oil well, that is, 20, 40, or possibly 80 acres. If offset locations to the discovery gas well are drilled in the same year as the discovery well, then additional reserves proved up by these offset wells will be put into the "discovery" category of reserves. It makes no difference how favorable the geological conditions are for the likelihood of more reserves being found beyond the limits of the discovery location (or offset drilled in the discovery year), such reserves are added by way of "extensions and revisions" when and if additional wells are drilled in later years.

In 1963, roughly 30 per cent of the total gas reserves added were from new discoveries. For the same year, only about 16 per cent of the total crude oil reserves added were from new discoveries. This reflects two things: (1) the acreage assigned to gas discovery wells is several times as large as for discovery oil wells; and (2) the initial recovery factor attributable to natural reservoir energies is several times as great for gas as for oil. Looking at the other side of the picture, one finds that extensions and revisions are much more significant for oil than for gas, for the same two reasons.

■ "Extensions and revisions" are defined in the AGA report in much the same manner as are discoveries:

> The reserves of all fields and pools are reviewed and revised upward or downward in each succeeding annual report to reflect additional information on preceding estimates. These changes are shown as "Extensions and Revisions."

As in the case of crude oil, extensions reflect primarily the acreage factor resulting from a growth in the areal extent of gas fields being drilled with development wells. Revisions reflect changes in the recovery factor resulting from added information gained by additional drilling or testing. Changes in the recovery factor caused by "secondary recovery" in the sense that the term is applied to crude oil do not occur.

The AGA does not report reserves attributable to extensions separately from those attributable to revisions even for the nation as a whole, while the API does make this distinction for crude oil. The question of why this lumping of revisions and extensions of gas reserves was done was explored with industry representatives to determine what, if any, reason there was for this aggregating. The reason may be that revisions make up a small part of the total extensions and revisions. This seems logical in the light of the high recovery factors usually found in gas reservoirs. It is argued that there is not a great deal of flexibility in the recovery factor if it is already in the 75 per cent range. On the other hand, the large acreage units assigned to gas wells introduce more uncertainty into initial reserves estimates and may result in the need for substantial upward or downward revisions at a later date.

Because of much wider gas-well spacing than oil-well spacing, gas fields may be drilled up or defined much more rapidly. If true, this means that gas reserves proved by extensions get into the total figure during a shorter time span, on the average, than do oil reserves. Added to this is the fact that gas reserves are rarely, if ever, added by something akin to the secondary recovery techniques used for oil, thus, the lengthy time span covered in proving up "secondary" oil reserves does not exist for gas.

One final comment should be made on reserves categories. The AGA does not report "discoveries" and "extensions and revisions" by type of reservoir—nonassociated, associated, and dissolved. Thus, it is impossible to determine whether or not gas added by "discoveries" is in one reservoir type or another. Similarly, the AGA does not report production by types of reservoirs, so that while we have year-end figures, for example, on non-associated gas reserves for successive years and know net changes in these reserves, we do not know whether the *gross* changes are large or small.

9

Estimating Procedures for Natural Gas Liquids Reserves

The estimation of NGL reserves is done primarily by the AGA Committee with assistance from the API Committee. Figures for NGL are found in both of these committees' reports, and these reserves are added to estimates of crude oil reserves to arrive at composite "liquid hydrocarbon" reserves for the nation and the individual states. Referring now to the present report, Table 3 shows the additions to NGL reserves at the end of 1964 and the amounts added during that year. On December 31, 1964, it was estimated that there were about 7.7 billion barrels of proved reserves of NGL in the United States. This is a little less than 20 per cent of the total proved liquid hydrocarbon reserve (crude oil plus NGL) for the nation. Table 2 (Chapter 2) presents a time series on NGL reserves since 1946, the year in which detailed estimates of NGL were begun. These figures will not be repeated here. Table 12 shows the geographic distribution of NGL reserves among the states as of the end of 1964. It should be noted that NGL reserves do not always occur in proportionate amounts with gas reserves. See, for example, the gas and NGL reserves figures for Kansas.

THE DEFINITION OF PROVED RESERVES

The AGA reports define proved NGL reserves in the following way:

Proved recoverable reserves of natural gas liquids are those contained in the recoverable gas reserves subject to being produced as

116

natural gas liquids by separators or extraction plants, now in operation, under construction or planned for the immediate future. For
purposes of developing reserve estimates, natural gas liquids are
defined as those hydrocarbon liquids which, in the reservoir, are
either gaseous or in solution with crude oil and which are recoverable
as liquids by the processes of condensation or absorption which take
place in field separators, scrubbers, gasoline plants, or cycling plants.
Natural gasoline, condensate, and liquified petroleum gases fall in
this category. While the liquids so collected and the products derived
from them in some of the modern plants are known by a variety of
names, they have been grouped together here under the general
heading "Natural Gas Liquids."

In the API report there is a footnote pertaining to NGL which reads
as follows:

> The API Committee includes in its crude oil figures all condensate
> which comes out of the separator with the crude oil and is run with
> the crude oil as part of the crude oil stream. All other condensate is
> included by the Natural Gas Reserves Committee in its figures on
> Natural Gas Liquids.

NGL reserves estimation presents a perplexing problem since these
liquids are in a sense an intermediate product—lighter than what is usually
considered crude oil and heavier than what is usually considered natural
gas. The definitions quoted above handle NGL in what is probably the
only practical manner. "Condensate"[1] which is trapped with the crude oil
in gas-oil separators cannot in any meaningful way be separated from the
oil, and thus these liquids are included in crude oil reserves. In a technical
sense they may be NGL but are mixed with crude to the extent that the
two become indistinguishable. The liquids produced in this manner might
better be referred to as "stock tank liquids" rather than crude oil. The
AGA views NGL as a product *contained* in the natural gas stream and
which is separable as a liquid from the gas stream after the gas is produced.
NGL and natural gas are perhaps best thought of as the two components
of "wet gas." This wet gas can be separated into "dry gas" and NGL.

[1] The term "condensate" causes considerable confusion in the industry and outside
of it. Technically, the term condensate should probably be restricted to the "retrograde
condensate" that is the product of true condensate reservoirs. "Field condensate" is a
term applied in the industry which applies to all liquids that come out of gas at the
wellhead. Such liquids are properly termed "natural gas liquids." Things are made
more confusing by the fact that a well may have as its major product true retrograde
condensate but also produce crude oil.

TABLE 12. ESTIMATED PROVED RECOVERABLE RESERVES OF NATURAL GAS LIQUIDS IN THE UNITED STATES [a]

(Thousands of barrels of 42 U. S. gallons)

	Reserves as of December 31, 1963 (1)	Changes in Reserves during 1964			Total (columns 6 + 7 + 8, also columns 1 + 2 + 3 less column 4) (5)	Reserves as of December 31, 1964		
		Extensions and revisions (2)	Discoveries of new fields and new pools in old fields (3)	Net production (4)		Non-associated (6)	Associated (7)	Dissolved (8)
Arkansas	18,695	1,871	41	1,597	19,010	2,025	8,631	8,354
California [c]	290,070	6,624	1,610	25,340	272,964	9,745	80,513	182,706
Colorado	21,990	5,812	1,189	3,263	25,728	3,750	2,710	19,268
Illinois	3,744	(—) 6	7	574	3,171	0	0	3,171
Indiana	95	13	0	17	91	3	3	85
Kansas	169,241	49,981	676	9,912	209,986	199,990	7,561	2,435
Kentucky	51,005	3,936	1,968	3,451	53,458	53,458 [b]	0	0
Louisiana [c]	1,840,823	170,782	59,878	129,983	1,941,500	1,681,799	197,543	62,158
Michigan	5,326	1,635	285	1,385	5,861	969	2,931	1,961
Mississippi	33,148	2,363	104	2,740	32,875	24,969	1,946	5,960
Montana	9,978	3,584	0	590	12,972	2,304	0	10,668
Nebraska	2,775	755	0	419	3,111	2,024	241	846
New Mexico	558,233	53,078	618	35,333	576,596	398,048	44,570	133,978
North Dakota	82,017	(—) 12,517	0	2,360	67,140	0	19,313	47,827
Ohio	0	0	1,200	66	1,134	0	0	1,134
Oklahoma	328,193	40,531	3,225	29,047	342,902	204,285	44,900	93,717
Pennsylvania	1,437	0	0	59	1,378	1,378 [b]	0	0
Texas [c]	4,042,358	117,602	73,318	273,490	3,959,788	2,089,045	633,807	1,236,936
Utah	46,591	7,970	2,137	2,041	54,657	613	20,000	34,044
West Virginia	68,779	5,617	2,808	7,430	69,774	69,774	0	0
Wyoming	99,480	(—) 1,929	1,978	6,993	92,536	47,654	379	44,503
TOTAL	7,673,978	457,702	151,042	536,090	7,746,632	4,791,833	1,065,048	1,889,751

[a] Includes condensate, natural gasoline and liquefied petroleum gas.
[b] Not allocated by types but occurring principally in the column shown.
[c] Includes offshore reserves.
SOURCE: Same as Table 1.

However, one immediately runs into the problem that "wet" and "dry" are relative terms, and a gas stream, after it has been processed, may have some "liquids" remaining.

The AGA Committee apparently follows a two-step process in estimating gas and NGL reserves. It first makes an estimate of "wet" gas reserves or gas and liquids combined. It then estimates the amounts of liquids contained in the gas in each specific reservoir. The gas figures are then reduced by the amount of shrinkage resulting from the removal of the liquids. The "dry" gas is what is reported in the annual report on gas reserves. The table in this report containing the state breakdown of gas reserves has the following footnote: "Excludes gas loss due to natural gas liquids recovery." This estimating procedure is complicated by the fact that NGL reserves are estimated only for those situations in which plants are available or planned to process gas for liquids removal. This is discussed next.

NGL figures are significantly influenced by the number and types of processing plants in operation, under construction, or definitely planned, since these reserves are included *only* if the liquids are recoverable by these plants. Raw-gas streams which contain NGL but which are used, flared, or vented in the raw state without first having the liquids removed do not contribute to the NGL reserves.[2] No doubt much of this gas contains some liquids which could be recovered if the gas were put through a cycling or processing plant. Presumably these liquids in unprocessed gas are included in gas reserves figures, while the liquids from gas reserves which go through , plants are included in NGL data. Any other method of estimation would result either in double counting or in the omission of some hydrocarbons from all the categories of reserves. As more plants are built and more raw gas put through them, NGL reserves should rise, even without the "discovery" of new gas reserves, and gas reserves should shrink. It is impossible to know what quantities are involved in this area, but it seems likely that they are rather small. Clearly, the AGA subcommittees have a considerable task in making adjustments in NGL and gas reserves data to reflect the construction of new processing plants or the putting of additional gas streams through existing plants.

[2] The Bureau of Mines reports that 1.9 trillion cubic feet of gas was consumed for "field use" in 1961. This is a "wet gas" figure which includes the gas volume for all natural gas liquids, including NGL recovered at plants from this gas. It is impossible to determine what part of the "field use" gas is put through plants or what quantity of NGL is included in the gas. *1961 Minerals Yearbook, Vol. II, Fuels* (Washington: Government Printing Office, 1962), pp. 328-29.

While "economic conditions" are not noted in the definition of NGL reserves, it is evident from the preceding discussion that these conditions must often dictate gas-oil separation and the construction of cycling and processing plants. NGL have grown in importance in postwar years. In 1947 NGL production accounted for about 8 per cent of total liquid hydrocarbon production; by 1964 this figure had risen to almost 17 per cent. Part of this rapid rise has been caused by the fact that crude oil production has been curtailed or kept static by state conservation agencies, or the market, while NGL production has been little affected by conservation regulations; partly it is due to the rise in gas production, and partly to the rapid construction of plants for removing liquids from gas.

Further discussion of the differences between nonassociated, associated, and dissolved gas reservoirs need not detain us here. What was said earlier about the gas reserves found in these categories of reservoirs applies equally to NGL reserves data. As was true in the case of gas reserves, it is impossible to know *gross* additions to NGL reserves by type of reservoir. This can only be determined if production by reservoir type is reported, and thus far these data are unavailable.

NGL reserves are treated in the same way as natural gas reserves as far as revisions and extensions are concerned. There is no national or state breakdown between reserves added by extensions and reserves added by revisions. The dependence of NGL reserves figures on (1) gas reserves figures and (2) NGL processing plant construction makes such a breakdown even more complicated than for gas or crude oil. Since revisions and extensions are reported on an aggregated basis, it seems likely that these data are built up from subcommittee estimates of each category for each producing area.

10

National Petroleum Council on Productive Capacity, Availability and Reserves of Natural Gas and NGL

For many years gas had the status of an unwanted stepchild of oil, and very little attention was paid, until fairly recently, to trends in gas reserves increments or flows over time, gas availability, or ability to produce gas. Since fuel availability over the long run is the information which is more vital to private and public planning than proved reserves data, it seems strange that more effort has not gone in this direction. In its 1952 Report, the National Petroleum Council did present a range of forecasts of gas availability for the years 1951 through 1955. In the words of the gas subcommittee of the NPC:

> The Subcommittee has deliberately used conservative estimates of availability of non-associated gas because the limiting factor on utilization of natural gas at the present time is still the pipeline facilities for transporting gas. The current availability of natural gas substantially exceeds the combined total of local consumption plus the capacity of existing large pipelines.
>
> Availability of associated [this included dissolved gas] gas was determined in relation to the availability of oil with which it would be produced.[1]

[1] National Petroleum Council, *Petroleum Productive Capacity*, 1952, p. 28.

ASSIGNMENT OF GAS RESERVES TO THE YEAR
OF DISCOVERY

The 1952 NPC study made no attempt to relate reserves to productive capacity or to date back gas reserves to years of discovery so as to establish a trend in additions to gas reserves over past years. It was not until 1961 that an example of this type of analysis was done for gas, appearing in the report of the NPC on *Proved Discoveries and Productive Capacity* discussed earlier in connection with crude oil. A similar analysis was included in the 1965 NPC report. While the 1945 Petroleum Administration for War study dated back crude oil to the year of discovery, this study had no similar analysis for gas. Thus, the recent NPC reports are the first instances of any serious effort to trace, historically, patterns of gas reserves development. The same is true for NGL reserves. Tables 13 and 14 show the summary of natural gas and NGL estimated discoveries assigned to fields discovered in the years 1919–1958, from the vantage point of 1963.

The previous discussion of the crude oil portions of the 1961 and 1965 NPC reports has relevance here as far as techniques and procedures are concerned and will not be repeated.[2] As in the case of crude oil, the NPC built upon reported proved reserves, using the American Gas Association definitions for reserves. Reserves were credited back to the year of discovery, by fields and reservoirs. The significant difference between crude oil and gas lies in the smaller proportion of ultimate reserves reported in the year of discovery for oil, as compared with gas. For example, "discovery" reserves assigned to crude oil fields in the period 1936–59 were about 10.92 billion barrels out of a total increment to reserves of 63.1 billion barrels. Thus, about 83 per cent of stated reserves of oil came from extensions and revisions. Between 1947 and 1959, out of a total increment to gas reserves of 216.8 trillion cubic feet, 149.5 trillion cubic feet or about 69 per cent were assigned to extensions and revisions. It was pointed out earlier in this study that the initial or discovery reserves estimates for gas may be higher than for oil, because the natural recovery forces are more efficient for gas, and because development of a field proceeds more rapidly for gas than for oil with the considerably wider gas-well spacing. Also, the additions to reserves of oil generated by such things as improvements in secondary recovery and pressure maintenance techniques have no counterpart in the gas reserves picture. This conjecture of greater gas discovery reserves seems to be borne out by the NPC data, although figures for comparable time periods are not available.

[2] See Chapter 3.

TABLE 13. SUMMARY OF ESTIMATED DISCOVERIES OF NATURAL GAS NOW ASSIGNED TO FIELDS DISCOVERED IN YEARS SHOWN

(Millions of cubic feet at 14.65 psia and 60° F)

Year	District 1 [a]	District 2	District 3	District 4	District 5	Total U.S.
1919 [b]	15,238,385	13,300,936	47,252,408	1,307,925	5,868,245	82,967,899
1920	365,421	1,445,231	87,462	37,567	1,095,818	3,031,499
1921	177,877	895,598	2,783,021	44,799	1,084,809	4,986,104
1922	98,322	28,373,183	5,361,533	239,887	131,370	34,204,295
1923	125,108	1,417,991	407,136	20,844	393,265	2,364,344
1924	74,288	653,245	664,978	41,656	313,800	1,747,967
1925	152,540	521,784	850,102	64,271	38,708	1,627,405
1926	396,515	1,368,540	2,324,582	681,105	264,400	5,035,142
1927	137,564	1,539,868	12,631,681	242,065	108,381	14,659,559
1928	755,015	2,405,712	1,207,825	240,109	2,011,692	6,620,353
1929	32,784	568,545	10,424,661	27,845	64,324	11,118,159
1930	228,835	600,121	5,936,368	189,355	60,332	7,015,011
1931	71,149	320,994	2,555,703	93,127	1,217,427	4,258,400
1932	106,924	237,042	3,183,276	133,029	470,181	4,130,452
1933	206,462	572,242	2,838,958	755,030	193,943	4,566,635
1934	57,180	526,466	6,015,224	175	495,496	7,094,541
1935	136,983	677,457	13,939,956	155,945	52,826	14,963,167
1936	906,722	244,653	17,525,286	1,823	4,401,502	23,079,986
1937	87,747	783,151	15,043,029	61,301	1,684,294	17,659,522
1938	51,368	591,647	12,328,673	911,723	1,166,063	15,049,474
1939	69,707	431,513	13,445,353	0	1,218,571	15,165,144
1940	65,196	792,894	10,851,760	2,127	103,583	11,815,560
1941	101,324	947,915	6,887,855	21,779	596,937	8,555,810
1942	115,868	442,338	6,328,416	86,919	119,987	7,093,528
1943	129,515	842,554	6,636,001	43,491	244,919	7,896,480
1944	133,570	411,811	7,674,735	223,431	298,973	8,742,520
1945	650,005	1,512,621	12,839,746	62,878	328,186	15,393,436
1946	155,137	2,109,884	5,028,005	597,380	192,932	8,083,338
1947	149,125	1,473,119	9,134,369	119,738	207,202	11,083,553
1948	158,099	252,075	7,802,345	427,521	191,210	8,831,250
1949	163,864	1,106,012	20,157,890	345,684	310,289	22,083,739
1950	208,143	466,111	7,331,874	501,209	267,082	8,774,419
1951	328,464	1,614,764	7,516,392	809,277	216,449	10,485,346
1952	118,174	4,555,908	10,546,440	172,631	381,868	15,775,021
1953	189,212	468,893	10,716,730	717,711	388,975	12,481,521
1954	165,148	2,002,931	11,210,288	432,551	184,034	13,994,952
1955	160,671	948,155	7,269,997	696,594	257,340	9,332,757
1956	134,222	1,606,402	7,528,548	595,831	347,675	10,212,678
1957	191,848	1,268,092	11,691,871	1,039,079	182,559	14,373,449
1958	247,064	979,917	8,776,286	653,398	526,137	11,182,802
Total	23,041,545	81,278,315	352,736,763	12,798,810	27,681,784	497,537,217

[a] For composition of districts, see Table 4, footnote a.
[b] 1919 includes all previous years.

SOURCE: National Petroleum Council. For full reference, see Table 4.

TABLE 14. SUMMARY OF ESTIMATED DISCOVERIES OF NATURAL GAS LIQUIDS NOW AS-
SIGNED TO FIELDS DISCOVERED IN YEARS SHOWN

(Thousands of barrels)

Year	District 1 [a]	District 2	District 3	District 4	District 5	Total U.S.
1919 [b]	111,073	408,796	778,347	42,726	310,082	1,651,024
1920	3,659	62,176	424	0	65,581	131,840
1921	1,667	34,278	151,087	0	76,210	263,242
1922	558	451,862	96,552	0	4,509	553,481
1923	879	47,685	10,427	980	18,486	78,457
1924	391	23,986	6,827	0	15,043	46,247
1925	1,260	5,769	8,221	0	1,489	16,739
1926	502	57,909	56,694	6,840	9,971	131,916
1927	1,258	56,733	307,765	283	5,303	371,342
1928	7,760	89,555	29,386	0	82,702	209,403
1929	228	11,646	247,636	0	1,509	261,019
1930	1,839	14,875	720,549	481	2,144	739,888
1931	160	3,080	100,568	1,200	61,891	166,899
1932	999	8,027	80,114	8,000	17,418	114,558
1933	152	20,996	28,207	19,100	2,552	71,007
1934	195	11,574	132,057	0	18,598	162,424
1935	975	19,276	475,148	10,688	573	506,660
1936	10,322	4,885	440,570	0	66,002	521,779
1937	184	29,720	427,060	0	81,569	538,533
1938	156	38,395	324,962	25,643	78,551	467,707
1939	546	16,356	392,127	0	113,923	522,952
1940	189	30,263	414,189	0	6,048	450,689
1941	439	45,652	189,747	0	13,959	249,797
1942	631	15,772	239,054	22,662	3,284	281,403
1943	641	33,920	171,911	0	3,200	209,672
1944	800	17,647	239,530	80	5,656	263,713
1945	8,356	27,738	432,524	8	11,341	479,967
1946	862	79,647	170,879	11,462	8,469	271,319
1947	1,687	39,650	274,742	120	2,492	318,691
1948	1,430	8,897	496,609	5,129	11,048	523,113
1949	981	29,497	527,275	3,994	22,065	583,812
1950	860	16,521	221,809	21,948	4,007	265,145
1951	777	40,978	255,617	22,924	2,055	322,351
1952	822	73,881	254,222	5,482	5,482	339,889
1953	762	18,383	393,536	22,631	2,231	437,543
1954	1,700	38,469	283,126	7,336	6,083	336,714
1955	1,353	36,430	209,283	3,423	8,770	259,259
1956	851	34,550	168,286	25,535	8,577	237,799
1957	956	29,508	348,170	24,042	1,742	404,418
1958	39,989	19,281	252,196	6,116	2,634	320,216
Total	208,849	2,054,263	10,357,433	298,833	1,163,249	14,082,627

[a] For composition of districts, see Table 4, footnote a.

[b] 1919 includes all previous years.

SOURCE: National Petroleum Council. For full reference, see Table 4.

In 1961 the NPC summarized its findings on gas and NGL historical data in the following way:

> It may be concluded from the preceding considerations that future additions to estimated recoveries of natural gas and of natural gas liquids may be expected from fields already discovered. As in the case of crude oil, therefore, the historical tabulations presented . . . should not be construed as representing the accurate record of discovery trends. Furthermore, the conclusions heretofore reached with respect to crude oil may be considered applicable also to natural gas and natural gas liquids.[3]

In the case of NGL, the story is almost identical to that of oil, that is, the "discovery" reserves estimates for NGL are about 16 per cent of total reserves assigned during the 1947–59 period; "extensions and revisions" are about 84 per cent. During this period, improved markets for NGL, improved technology and facilities for recovering and processing the liquids, and the conservation prod wielded by the state governments all had a part in boosting extensions and revisions of gas liquids.

Since the Petroleum Administration for War report of 1945 had no analysis of gas data, the NPC Report cannot make the interesting temporal comparisons that were made for crude oil (see Chapter 3 of this study). No doubt the Department of the Interior feels, as evidenced by its request for the 1961 NPC Report, that gas is now of sufficient importance in the total energy picture to warrant the more extended treatment afforded oil in past reports. This seems appropriate since consumption of natural gas in the United States now accounts for about 29 per cent of total energy requirements compared to about 42 per cent for oil and oil products (computed on a heat equivalent basis).

PRODUCTIVE CAPACITY OF NATURAL GAS

Productive capacity of natural gas and NGL is estimated in the NPC study in much the same manner as productive capacity for crude oil. Some industry people feel that productive capacity is more difficult to define for gas and NGL than for oil, and that estimates of capacity are more difficult to make. Since gas occurs in nature either by itself or with crude oil and may, or may not, contain NGL, productive capacity for gas must be

[3] National Petroleum Council, *Proved Discoveries and Productive Capacity* (Washington: 1961), p. 24.

computed for each type of reservoir—nonassociated, associated, and dissolved. Table 15 shows the NPC estimates of productive capacity of gas by Petroleum Administration for Defense (PAD) district and by type of reservoir as of January 1, 1964. The total productive capacity of gas reported by the NPC assumes that oil is produced at productive capacity. As we noted earlier, oil is not produced at capacity in the United States and, barring some emergency, probably will not be produced at these full rates in the near future. The NPC points out that the concept of productive capacity ". . . reflects the capacity of gas and oil reservoirs and is a measure of the amount that could be produced from existing wells without regard to limitations of markets, transportation, and processing facilities." It is ". . . in no sense a measure of current availability of gas for consumption, and it should not be so construed. . . . Any increase in daily production would require adequate markets and the installation of additional pipeline, compressor, and other facilities. . . ."

TABLE 15. MAXIMUM PRODUCTIVE CAPACITY OF NATURAL GAS IN THE UNITED STATES [a]

(Thousand cubic feet daily at 14.65 psia and 60° F)

District [b]	January 1, 1964				January 1, 1960 Total
	Non-associated	Asso-ciated	Dis-solved	Total	
1. (East Coast)	919,000	0	11,200	930,200	990,300
2. (Mid-Continent)	12,176,300	1,083,600	1,448,100	14,708,000	8,203,600
3. (Gulf Coast)	58,828,300	5,532,000	12,963,700	77,324,000	58,121,200
4. (Rocky Mountain)	1,789,000	140,400	299,500	2,228,900	2,327,000
5. (West Coast)	1,217,000	0	1,176,000	2,393,000	1,862,000
Total	74,929,600	6,756,000	15,898,500	97,584,100	71,504,100

[a] This statement does *not* include any gas available from storage reservoirs. It *does* include gas available from reservoirs being cycled. In some cases, associated gas production cannot be distinguished from dissolved gas production. In these cases, all such gas is tabulated as dissolved.

[b] For composition of districts, see Table 4, footnote a.

SOURCE: National Petroleum Council. For full reference, see Table 4.

The NPC estimates that of the total capacity of 97,584 million cubic feet per day (MMCFD) in 1964, 74,930 MMCFD was in nonassociated gas reservoirs, 15,898 MMCFD was dissolved in oil, and 6,756 MMCFD associated with oil. We know that total gas production in 1963 averaged over 41,400 MMCFD, but we do not know production by type of reservoir. Without this information it is impossible to estimate whether the nonassociated or the dissolved and associated reservoirs have the greatest "excess" producing capacity. Since oil is not currently being produced at its producing capacity, all the dissolved and associated gas capacity is not being used.

It should be noted in passing that there is an area of possible confusion concerning gas capacity in associated gas reservoirs. In gas cap reservoirs the production of oil at efficient capacity usually requires that gas production be kept quite low, relative to the gas potential. If the gas cap gas were produced at its capacity, and if oil production were disregarded, the gas producing capacity of these reservoirs would be substantially enhanced. The NPC does not define what it means by "productive capacity," nor is there discussion of M.E.R.'s (maximum efficient rates of production) for oil or gas as this concept relates to productive capacity. We would assume, given the silence in the NPC report on this matter, that oil capacity takes precedence. In instances in which oil recovery must suffer if gas producing rates are ignored, presumably the recorded producing capacity for gas is that rate at which oil can be recovered efficiently.

Figures on productive capacity for gas include the capacity of shut in fields and of gas currently being cycled. They exclude the capacity of gas storage fields.

The NPC estimates that during 1963 the *peak day* gas requirement for the entire nation was 56,292 MMCFD. Thus the nonassociated gas producing capacity of 74,930 MMCFD exceeded the total peak day requirement by 18,638 MMCFD. This indicates that there is likely a large "excess" or unused capacity in all types of gas reservoirs.

PRODUCTIVE CAPACITY OF NATURAL GAS LIQUIDS

The productive capacity of NGL is directly a function of gas production. In earlier reports on NGL availability the NPC had estimated NGL capacity on the basis of existing plants and equipment. The 1961 and 1965 NPC reports have used another method. The capacity figures are based on the NGL that *could* be recovered if oil and gas were both produced at capacity and if all NGL were removed. Thus for January 1, 1964, productive capacity was about 2,802,700 barrels daily, while NGL production during 1963 was only about 1,413,000 barrels daily. The productive capacity of NGL for the United States by PAD district, by type of reservoir, is shown in Table 16.

TABLE 16. MAXIMUM PRODUCTIVE CAPACITY OF NATURAL GAS LIQUIDS IN THE UNITED STATES

(Thousands of barrels daily)

District [a]	January 1, 1964				January 1, 1960 Total
	Non-associated	Associated	Dissolved	Total	
1. (East Coast)	22.3	0	0	22.3	10.2
2. (Mid-Continent)	92.4	23.1	53.7	169.2	162.6
3. (Gulf Coast)	1,564.2	219.2	701.8	2,485.2	1,501.8
4. (Rocky Mountain)	16.0	1.3	26.7	44.0	37.0
5. (West Coast)	3.0	0	79.0	82.0	88.0
Total	1,697.9	243.6	861.2	2,802.7	1,799.6

[a] For composition of districts, see Table 4, footnote a.

SOURCE: National Petroleum Council. For full reference, see Table 4.

Earlier NPC reports on NGL productive capacity include only those amounts which are recoverable with existing facilities. The figures from these reports are shown in Table 17. It is unfortunate that the NPC has changed so drastically the definition of productive capacity as it applies to NGL. Clearly, the new definition has certain advantages and more closely conforms to the capacity concept as applied to oil and gas. However, the "availability" concept is quite useful also and perhaps should be continued so as to run parallel with the capacity data developed under the new definition.

For a more detailed analysis of the NPC techniques as well as the limitations of and problems involved in using the NPC data, the reader should consult the discussion given earlier for crude oil (see Chapter 3).

TABLE 17. U.S. AVAILABILITY OR PRODUCTIVE CAPACITY OF NATURAL GAS LIQUIDS, 1951-65 [a]

(Thousands of barrels daily)

1951—Jan.	573
1953—Jan.	694
1954—July	765
1957—Jan.	845
1960—Jan.	1,800 [b]
1964—Jan.	2,803 [b]

[a] Availability concept is used for 1951, 1953, 1954, and 1957. Productive capacity concept used for 1960 and 1964.

[b] Reported on a different basis. Includes all recoverable liquids in all potentially producible oil and gas.

SOURCES: Reports of the National Petroleum Council for 1951, 1953, 1955, 1957, 1961, and 1965.

11

Extending the Boundaries of
Reserves Estimates for Gas

As one proceeds to broaden the definition of "reserves" of petroleum, moving from the restricted "proved" reserves concept of the API and the AGA through "probable" reserves, "possible" reserves, and finally "ultimate" reserves, the preciseness of the estimates becomes less certain, and the quantities become increasingly speculative orders of magnitude. For crude oil, the first step away from the proved reserves or working inventory concept was that of estimating oil recoverable from known fields (1) using existing secondary recovery techniques and (2) using secondary recovery techniques on the immediate technological horizon. Such estimates, made by Paul Torrey and his IOCC Committee for oil, were discussed earlier. Since gas resources are almost entirely recoverable using primary methods, there are no secondary gas reserves comparable to the IOCC estimates for oil.

From the earlier discussion of the API and AGA definitions of crude oil reserves as compared to natural gas reserves, it is clear that "proved" oil reserves have a stated, albeit rather fuzzy, economic parameter, while "proved" gas reserves do not. As we also indicated earlier, this means that some part—and we are unable to judge how much—of the reported proved gas reserves are currently noneconomic and thus more closely akin to the "probable" or "possible" categories of crude oil reserves. These noneconomic reserves cannot be precisely categorized, but they should not be left under the "proved" definition if this definition includes economic criteria. The recent work of the Society of Petroleum Engineers discussed

129

earlier (see Chapter 5) has apparently run head on into this problem. The SPE Committee is interested in standardizing definitions of reserves categories for property valuation purposes and has worked toward this end within the confines of the API proved oil reserves concept. The API definition for crude oil has an economic parameter and thus provides a workable base for the SPE Committee's work. The AGA has not yet been able to resolve this difficulty with the SPE of not having an economic parameter, and thus apparently the two groups have no common basis upon which to discuss their mutual problems. If, as seems likely, the API definition of proved crude oil reserves is, in effect, now spelled out by the SPE with greater emphasis on economics, the little-noticed difference between oil and gas reserves definitions will become glaring.

A second step away from the proved reserves concept of the AGA is to estimate probable reserves in fields which have been discovered but have not been partially or fully developed. From discussions we have had with industry representatives, there seems to be little doubt that individual oil and gas producing companies do make estimates of the total quantities of gas that can be recovered from newly discovered but as yet undeveloped or only partially developed fields. Such estimates are based upon what reservoir information and productive history, if any, is available from completed wells and upon the rough outlines of the reservoir as determined by geological and geophysical information. In the discussion of company estimates of crude oil reserves, we have outlined the procedure followed and the definitions of reserves categories used by companies. What was said there with reference to oil reserves is generally applicable to gas reserves (see Chapter 5).

There is one important difference between oil and gas which should be re-emphasized. Well-spacing for drilling gas wells is considerably wider than for oil wells, and the acreage assigned to a well as proven acreage is considerably greater. Thus gas reserves tend to be "proved up" faster than oil reserves, or, put in other terms, a gas field is usually defined more rapidly and with fewer wells drilled, particularly if there is a market for the gas. This being the case, gas reserves move more quickly from the "probable" to the "proved" category on companies' books.

The techniques proposed by J. R. Arrington (see Chapter 5) for estimating the eventual size of crude oil reserves based on new discoveries are not limited to oil; they are equally applicable to gas reserves, given the data for trend analysis which the method requires.

The approach proposed by A. D. Zapp for estimating reserves of crude oil from productive capacity (see Chapter 6) probably has little applicability to gas. His study does, however, make a significant point bearing on gas reserves. With a growing demand for gas, it may become economically feasible to recover more of the gas that is in place than is currently being done. Gas wells today are usually abandoned when the reservoir pressure declines to a point at which insufficient gas comes from the well to make it profitable to operate. If gas gets more valuable, abandonment pressures will drop. In certain high price gas areas, it may even be profitable to install vacuum pumps and suck gas from the ground.[1] Economic factors control in such a situation. If in 1945 some gas fields were approaching abandonment with an average wellhead price of 4.9 cents per mcf (thousand cubic feet), one might raise the question of what happened to abandonment pressures during the late 1940's and 1950's when, by 1960, average wellhead prices reached 14.0 cents per mcf. If abandonment pressures went down, reserves must have gone up. To what extent AGA reserves estimates have taken this into account is not known; the Reserves Committee is silent on the point.

In one sense, the question of abandonment pressures is rather academic. In an interesting article, D. T. MacRoberts makes the following point:

> As one searches the records there is one especially impressive fact. With the possible exception of the eastern gas fields, although they appear to have been numerous and small, there are very few major gas fields that have been abandoned. . . . I can't recall a single field that has produced a trillion cu. ft. and which has been abandoned. The major gas fields of the southwestern United States are still producing.[2]

MacRoberts notes that the Monroe Field in northern Louisiana has been producing gas at pressures of 25 lbs. for 20 years and still has some life ahead. The point to be made is that abandonment is difficult to predict even for an operator vitally interested in the future of a field. It is even more difficult for a group such as the AGA reserves committee to predict. Thus, while it is true that economic forces are significant in relation to the abandonment of fields, such forces are extremely difficult to isolate and account for specifically. One gets the feeling they are there but in rather shadowy, uncertain form.

[1] A. D. Zapp, *Geological Survey Bulletin 1142-H* (Washington: Government Printing Office, 1962), p. H-19.

[2] "Abandonment Pressure of Gas Wells," *1962 Symposium on Petroleum Economics and Valuation,* Society of Petroleum Engineers, Dallas Section, 1963, p. 58.

In general, as we have shown, new discoveries of gas are translated into proved reserves through revisions and extensions more rapidly than in the case of crude oil. There is therefore less room for speculation concerning the amount of future additions to reserves in known fields, over and above current proved reserves. Moreover, since the usual recovery rate is already high, there is less room for the kind of speculation which arises in the case of crude oil, about additions to reserves which may come from improvement in recovery methods. For these reasons, the proved reserves estimates for gas are presumably considerably closer to reasonable expectations for ultimate recovery from known fields than are the proved reserves estimates for crude oil, at any point in time.

SOME SPECIAL PROBLEMS

This discussion of gas reserves has alluded to a number of unique problems that pertain to gas because of its physical occurrence, its regulatory status, or its special institutional surroundings. These problems are summarized here.

There is considerable need for more detailed reporting of additions to gas reserves and production by type of reservoir—nonassociated, associated or dissolved. Similarly, and perhaps a derivative, there is a need for productive capacity and excess productive capacity information by type of reservoir. The Federal Power Commission has the task of setting field prices for gas or, in effect, selecting a price which equates supply and demand. The supply of gas comes from gas wells and from oil wells scattered across the country. There is no information readily available which reveals whether the available gas supplies are nonassociated, associated or dissolved, or where in the country various amounts of each type are located.

Closely related to this is another problem with a regulatory aspect. Gas supplies (proved reserves) are usually sold under a long-term contract. Frequently the contract commits all the gas reserves under a specified lease to the buyer for as long as gas can economically be produced from that lease. The amount, type and location of available gas supplies, that is, those reserves not already committed to future uses under long-term contracts and are therefore marketable, are of considerable importance to gas buyers and sellers as well as to the FPC. Some reporting technique might be devised which reveals this information.

As a corollary to gas reserves information of the type outlined here, there is a need for similar data on gas deliverability. Because gas moves only by pipeline and because uninterrupted minimum flows are essential, deliverability becomes a critical question. Information is needed on deliverability of gas produced jointly with oil and thus subject to some extent to oil conservation regulation.

Finally, there is a need to have annual *gross* extensions, revisions, and discoveries of gas reserves reported separately by states for nonassociated, associated, and dissolved gas reservoirs. Such information must be available in the working papers of the AGA reserve subcommittees. The collection and reporting of these data would not involve new methods but merely the additional work of recording the details from which the reported aggregate data are built.

With respect to natural gas liquids, there needs to be some thought given to reporting all NGL carried in all gas streams, whether or not the NGL is removed from the streams. It was noted earlier that such a task would be difficult, but some idea of the magnitudes of the quantities involved would be helpful. A more detailed reporting of NGL reserves data *vis-à-vis* extensions, revisions and discoveries of nonassociated, associated, and dissolved NGL would also add significantly to the information necessary to analyze fully our energy resources.

Much of the information outlined above probably exists in one form or another in company files. The FPC has proposed questionnaires which will bring forth some of it. Data collection can be extremely expensive to those furnishing the information, and there is clearly a need for discussion within the industry and between the industry and the state and federal bodies charged with implementing various aspects of our national energy policy as to what is needed, how practicable the collection is, and what costs must be incurred to bring forth what types of data. The need for better data is recognized in a general way by many industry and government people. There is, however, disagreement on the significance, on the details, and on the costs involved. It is impossible to ignore the plea of the industry that it is in business to make money, and that data which are of no benefit to the companies in the industry, either individually or collectively, must be collected under duress. Those proposing additional data (and we have suggested certain things which seem to us to be desirable for intelligent public policy in the fuels area) must justify their proposals. This is no easy task. We shall return to this subject in the last section of the study.

"ULTIMATE" GAS RESERVES

The urge to probe the unknown and to estimate "ultimate" gas reserves has grown stronger in recent years with the growing importance of gas in the national fuel supply picture. Interest has been generated by government studies on oil imports, by Federal Power Commission stress on long-run supplies for consumers, by the financial interests who have a stake in gas transmission lines, and by producers who are searching for ways to supplement their somewhat stagnant domestic oil revenues.

As was true in the case of "ultimate reserves" of oil, the term "reserves" should probably not be used in this context at all. Here we are discussing gas deposits that may, in part, exist in as yet undiscovered fields. To bring these deposits into the proved or probable reserves categories requires that the fields be found and developed.

October of 1964 brought the first "industry" effort that went beyond reporting proved gas reserves and delved into the realms of probable, possible, and ultimate. In its statement of scope and purpose, the newly formed Future Gas Supply Committee noted:

> The leaders of the natural gas industry have expressed a strong desire for a continuing long-range study of the requirements and supply of the natural gas in the United States. Many have made predictions concerning the gas requirements and the gas supply of the future. None has made a thorough nationwide study using the efforts and judgement of competent representation from all phases of the industry. The time has now come for industry to appraise its own future.[3]

The Committee proceeded to make estimates as of December 31, 1963, of "future available supplies of natural gas . . . [which] includes all of the natural gas in any particular area which it is believed could be found and produced under prevailing technical conditions, but over and above the estimated volume of proved recoverable reserves made by the American Gas Association Committee on Natural Gas Reserves."[4] This figure for the nation was 629.5 trillion cubic feet and is in addition to the 276.15 trillion cubic feet of reported proved reserves. The total future supply, therefore, is about 905 trillion cubic feet. The U. S. figure is broken down

---

[3] Future Gas Supply Committee, *Future Natural Gas Supply of the United States*, Vol. 1 (Omaha: 1964), p. 2.
[4] *Ibid.*, p. 9.

regionally between the "East," "Central," and "West," with the Central area accounting for 442.5 trillion cubic feet of the 629.5 figure. The Committee hastens to point out that "the 'future available supply' and the rate at which it can be developed is a function of economic factors and technologic improvements."

It is clear from the text that the Committee intends to issue a continuing series of reports on both future supplies and future requirements.[5] The procedures noted generally follow those of the API and AGA described earlier, at least in terms of a parent committee with areas of the country assigned to subcommittees for reporting purposes. The individual subcommittee member retains his own confidential working papers on his assigned area and reports only final estimates to his subcommittee chairman. Thus, as is the case for the API and AGA committees, there is substantial reliance on the expertise of individual members.

In discussing its definition of future supply the Committee makes the following comment:

> The technical approach to determining the future available gas supply of the United States was studied in three (3) categories— probable future supply, possible future supply, and suspected future supply. These are all inferred available future supplies and cannot be classed in any sense as reserves nor should they be confused with or referred to as such. After considerable discussion by the Committee, it was decided that in this initial report, the total future available natural gas supply of the United States would be reported as one figure—and not set out by the volume determined to be estimated for each category until such time as methods can be refined and agreement on terminology can be reached.
>
> However, the Committee is now working on improving its technical approach so that it will be possible to use these or similar categories for future studies and plans to make future reports on the basis of two or three categories of reasonable expectancy.[6]

This new report is obviously breaking new ground for the industry in providing reserves information beyond the "proved" category. Our discussions with industry representatives revealed a sharp division of opinion within the industry on the advisability of moving beyond proved reserves

---

[5] In December 1964, the Future Gas Requirements Committee issued its first *Report on Future Natural Gas Requirements of the United States*. A parent Future Gas Requirements and Supply Committee apparently oversees these two "super subcommittees."

[6] Future Gas Supply Committee, *op. cit.*, pp. 9-10.

reporting.[7] It is significant, perhaps, that this report did not come out under the name of the AGA even though the Association had a hand in its initial conception and, in fact, has an AGA staff member on the parent Requirements and Supply Committee. At least one significant barrier to going beyond the proved concept has been hurdled. An important and influential group in the oil and gas industry feels that information on "future" gas reserves is needed and that the definitional and technical problems of gathering the data, while difficult, can be solved. There is no mention of the Future Supply and AGA Gas Reserves Committees working together, but it certainly seems appropriate and, in fact, necessary that they do so, particularly for the estimation of what we have called reserves in undrilled portions of known fields. The Future Supply Committee and its subcommittees are fortunate to have the assistance and co-operation of competent technical people from major oil and gas producing companies and gas transmission companies. Hopefully, this co-operation from within the industry will expand. Such effort should enable the Committee to develop appropriate definitions for the concepts discussed in its first report, and also enable it to report future supplies data in greater geographical detail.[8]

The figures of the Future Supply Committee are by no means the first efforts at estimating "ultimate" gas reserves. The literature up to 1959 on ultimate gas reserves has been carefully reviewed by Netschert in *Energy in the American Economy*,[9] and the reader should consult this source for a discussion of these estimates. The range of reserves yet to be discovered as of January 1, 1960 is reported by Netschert as being from 247 trillion cubic feet to 937 trillion cubic feet.[10] When added to proved reserves of this date, the ultimate future supply ranges from 510 trillion to 1,200 trillion cubic feet. Mr. Netschert takes as his own estimate the upper limit of this range, or 1,200 trillion cubic feet.

Since the Netschert study there have been a number of other estimates of ultimate gas reserves. Mr. Zapp has estimated ultimate gas reserves at

[7] Mr. Muskat, chairman of the API Reserves Committee, viewed such extensions with skepticism because of definitional problems, although he felt that if these could be overcome such data proposals ". . . would no doubt merit serious consideration." Morris Muskat, "The Proved Crude Oil Reserves of the U. S.," *Journal of Petroleum Technology* (September 1963), p. 918.

[8] In May 1965, the Future Gas Requirements and Supply Committee announced that the Colorado School of Mines would sponsor this group.

[9] Sam H. Schurr and Bruce C. Netschert, *Energy in the American Economy, 1850-1975* (Baltimore: Johns Hopkins Press for Resources for the Future, Inc., 1960).

[10] *Ibid.*, p. 393.

2,650 trillion cubic feet.[11] This is perhaps one of the highest estimates ever made. A less well known estimate by the U.S. Department of the Interior in 1960 for the Joint Committee on Atomic Energy carries a figure of 3,000 to 5,000 trillion cubic feet of gas "in the ground in undiscovered deposits" in the United States and Canada combined.[12] In a Geological Survey Bulletin in 1961, Paul Averitt estimates ultimate reserves of natural gas at about 2,000 trillion cubic feet.[13] In a much lower range, Mr. Hubbert in 1962 estimated that the "ultimate potential reserves of natural gas" are in the range of 958 trillion to 1,053 trillion cubic feet.[14] Ralph L. Miller, geologist for the U.S. Geological Survey, estimated in late 1958 that ultimate natural gas reserves (less production to January 1, 1957) will be about 1,000 trillion cubic feet.[15] The *Report of the National Fuels and Energy Study Group* estimated in 1962 that natural gas reserves "recoverable from extensions to present areas and from pools yet to be discovered" are about 1,000 trillion cubic feet.[16] In 1963, D. C. Duncan and V. E. McKelvey of the U.S. Geological Survey made the following estimates: 268 trillion cubic feet of "known recoverable reserves"; 1,200 trillion cubic feet of "undiscovered recoverable resources"; and 850 trillion cubic feet of "undiscovered marginal resources" (not currently economic to produce if discovered).[17] This would total to 1,468 trillion cubic feet of recoverable gas plus the 850 trillion cubic feet of noneconomic gas. Finally a Resources for the Future Study by Landsberg, Fischman and Fisher estimates that ultimate gas reserves will range from 1,200 trillion to 1,700 trillion cubic feet.[18]

[11] As cited by M. King Hubbert in *Energy Resources, A Report to the Committee on Natural Resources of the National Academy of Sciences-National Research Council*, Publication 1000-D (Washington: National Academy of Sciences-National Research Council, 1962), p. 77.

[12] *Background Material for the Review of the International Atomic Policies and Programs of the United States*, report to the Joint Committee on Atomic Energy, Vol. 4, Joint Committee Print, 86 Cong., 2 sess. (Washington: Government Printing Office, 1960), p. 1529.

[13] *Coal Reserves of the United States—A Progress Report, January 1, 1960*, U.S. Geological Survey Bulletin 1136 (Washington: Government Printing Office, 1961), p. 101.

[14] Hubbert, *op. cit.*, p. 80.

[15] *Oil and Gas Journal*, September 1958, p. 136; and October 1958, p. 222.

[16] *Report of the National Fuels and Energy Study Group on An Assessment of Available Information on Energy in the United States*, to the Committee on Interior and Insular Affair , U.S. Senate, Committee Print, 87 Cong., 2 sess. (Washington: Government Printing Office, 1962), p. 79.

[17] Office of Science and Technology, *Research and Development on Natural Resources*, Report by the Committee on Natural Resources to the Federal Council for Science and Technology (Washington: Government Printing Office, 1963), pp. 43-44.

[18] Hans H. Landsberg, Leonard L. Fischman, and Joseph L. Fisher, *Resources in America's Future* (Baltimore: Johns Hopkins Press for Resources for the Future, Inc., 1963), p. 407.

The estimating technique for ultimate gas reserves usually involves two steps: (1) an estimation of ultimate recoverable oil reserves or oil in place; and (2) an estimation of a ratio of gas reserves to oil reserves yet to be discovered. This means that such gas estimates are subject to all the uncertainties and infirmities that are found in determining future ultimate oil reserves, plus the additional uncertainty of assuming some ratio of gas reserves to oil reserves for the future. As might be expected, having two major variables results in a wide range of estimates.

Much has been written on the clearly perceptible trend of a rising ratio in proved gas to oil reserves and in gas to oil production data. Much of this increase is attributed to deeper producing depths, at which depths gas is more likely to be found than oil. Less has been said about the economic incentives (that may exist or be forthcoming) to search for gas instead of oil and the possible influence of these incentives on the ratio of gas to oil in proved reserves. It seems likely that economic incentives have also been a factor in the rising ratio. At any rate, forecasters' sights have been progressively raised with respect to the appropriate ratio. As recently as 1950 an assumed ratio of 6,000 cubic feet of gas reserves per barrel of oil reserves was considered to be reasonable. This has been raised to 7,000 cubic feet per barrel by some and even 8,000 cubic feet per barrel by others.[19]

The important thing to note in these estimates of "ultimate" gas reserves is their general order of magnitude and direction. During the postwar years the magnitudes have roughly doubled—from about 750 trillion cubic feet to 1,500 trillion cubic feet; the direction has been consistently upward. Whether or not reserves of these magnitudes will be forthcoming and what the time distribution of development will be depends upon many factors, some of which are highly uncertain. Among the more important variables are: (1) the price incentives extant in the future for gas; (2) the trend in the search for new domestic reserves as between oil and gas; (3) the demands for gas relative to competitive energy supplies; (4) the developing technology of supplying gas and liquid hydrocarbon fuels from such things as oil shale, coal, and tar sands; and (5) the future role of oil and liquid methane imports in the domestic fuel market.

In some degree the range from optimism to pessimism is related to views upon the prospects for deep drilling. Mr. Zapp, for example, concludes that, because more gas is found at greater depths and because deeper drilling is a certainty, the outlook for gas reserves is good and certainly

---

[19] Hubbert assumes a ratio of from 6,250 cubic feet to 7,500 cubic feet per barrel; Davis assumes a ratio of 7,000 cubic feet per barrel; Landsberg, Fischman, and Fisher assume 8,000 cubic feet per barrel for the higher of their two estimates.

more favorable than for oil. Others, for example H. K. Hudson,[20] point to evidence that oil occurrence at great depths is likely to be small, inhibiting drilling where only gas is likely to be found; and also that the areal extent of sedimentary beds is much less extensive at greater depths and that individual reservoirs are smaller. These differences of analysis present an additional hazard in the speculation upon future discoveries of gas.

A final comment should perhaps be made on "in place" estimates for gas. Earlier discussion emphasized that estimates of oil in place are urged by some so as to get some idea of the amount of oil already discovered but not currently recoverable. We have such estimates in the IOCC reports on reserves Interstate Oil Compact Commission (see Chapter 4). This problem is much less critical for gas since the proved reserves of gas usually represent a large part of the gas in place. The opportunities for "secondary recovery" of gas seem rather remote at this time, and therefore figures on gas in place are perhaps not particularly important.

CLASSIFICATION OF GAS RESERVES IN THE SOVIET UNION

Two studies, one by Professor M. A. Adelman, *The Supply and Price of Natural Gas*,[21] and one by D. B. Shimkin, *The Soviet Mineral Fuels Industries, 1928–1958: A Statistical Survey*,[22] contain extremely interesting sections which describe the techniques used in the Soviet Union for classifying natural gas reserves. It is noteworthy for our purposes because it is the only systematic treatment, as far as we know, of the whole spectrum of probabilities of reserves. According to Adelman, citing a Soviet document, the Soviet system classifies gas reserves in the following way:

A_1—Reserves being currently exploited.

A_2—Reserves "tested in detail by bore holes."

B —"Reserves established quantitatively, sufficiently accurately by prospecting. The form of the body or the distribution of the natural type of useful mineral or the technology of processing are insufficiently known."

[20] H. K. Hudson, "Is the 'song of plenty' a siren song?" *Oil and Gas Journal*, June 17, 1963.

[21] Supplement to *The Journal of Industrial Economics*, 1962, pp. 113ff.

[22] Bureau of the Census, U.S. Department of Commerce, International Population Statistics Reports, Series P-90, No. 19 (Washington: Government Printing Office, 1962), pp. 153-67.

$A_1 + A_2 + B$ = Industrial reserves.

C_1—"Assumed reserves adjacent to explored portions outside contour [structure?] of a higher category, . . . estimated on the basis of geological study of natural and sparse artificial exposures and geophysical data," and "poorly prospective reserves."

C_2—Reserves "determined by geological premises."

Prospective Reserves—"Reserves potentially present . . . whose discovery can be anticipated in planning exploratory and prospecting work. The accuracy with which reserves of this category are determined is of course "less than that of category C_2 . . ." and the estimates "should be regarded as minimal values which will increase. . . . "

Professor Adelman points out that the concept of "industrial reserves" is somewhat broader than the AGA "proved reserve" concept but would include all proved reserves. In addition, it may include reserves which, in this country, are in known fields but which lie in undefined portions of fields and thus are excluded from the AGA "proved reserves" concept. There is some uncertainty as to the precise matching with United States concepts. The C_2 group of reserves is apparently meant to be more than a wild guess at "possible" reserves and includes those fields in which exploratory work is being done. As reserves in a field become better known and more fully developed, they are shifted from a lower classification to a higher one. Professor Adelman concludes by saying:

> . . . something of this kind would be very useful for other petroleum areas. . . . The Soviet system in effect provides for an annual estimate of the less certain reserves, and like every periodic system, including the API-AGA, repetition gives both increasing precision of the data and more insight into their meaning. A project under international auspices for the standardization and improvement of petroleum reserve statistics, ought to be of great value to all concerned.[23]

It seems to us that those groups in this country interested in improving reserves data might well consider Professor Adelman's suggestion and at least explore the reserves estimating techniques and categories used by the Soviet oil industry. If it is feasible, and this is a big "if," a time series of such data with annual descriptions of the movement of reserves from one category to another would provide immensely more insight into the United

[23] Adelman, *op. cit.*, p. 116.

States petroleum reserve position than the currently published reports on proved reserves of the API and AGA and the supplementary data provided by the NPC. Such a system may have substantial shortcomings and may involve prohibitive costs. The greatest inhibiting factor, already mentioned with reference to oil and also applicable for gas, is the fact that individual companies do not maintain records which could be aggregated in this way.

GAS RESERVES AND GAS DELIVERABILITY

While gas reserves estimates indicate the proved volumes of gas recoverable from known reservoirs, such figures tell nothing about the time distribution of production, which for many company and public policy decisions is a far more important piece of information. Gas "deliverability" for a given well or reservoir has been defined by one expert as "the volume of gas to be produced . . . during a given period of time, at a specified contractual rate of flow, against a certain well-head pressure, under prevailing physical conditions of the reservoir and the status of well development in the field, taking into account any other contractual and applicable regulatory restrictions."[24] It will be noted that there are three separate but related types of constraints on deliverability: (1) the physical constraints of the amount of gas in place and its characteristics, the reservoir character, and the extent and type of development in the field; (2) the private contractual constraints which may limit deliverability below what optimally could be physically produced; and (3) the regulatory constraints that may also limit deliverability below what could be physically produced and in some cases may limit the production agreed upon contractually by the buying and selling parties.

Estimates of gas deliverability are essential for company purposes in making revenue projections, which in turn are critical for short-run and long-run operating and investment expenditure planning. In many instances the forecasting of cash flows is what is of critical importance to internal company decision-making. Estimates of gas deliverability are also essential for public policy making in the areas of gas distribution, transmission, and production, although in this context the term "availability" might be more appropriate. For example, in arranging financing for either a new gas distribution system or a gas pipeline or a major expansion to

---

[24] Federal Power Commission, Area Rate Proceeding, Docket No. AR 61-1 (Permian Basin), Testimony of Ianel I. Gardescu, Transcript pp. 272-73.

either one, the regulatory authority must decide whether or not gas deliveries are sufficient to support the estimated markets for gas at the indicated selling prices. The failure of deliveries to do this may bring forth insufficient revenues and the resultant failure to recover fully the capital loaned to the company by bondholders. In such situations, gas deliverability and availability is clearly a question in which the public interest is involved from both the consumers' standpoint and the investors' standpoint.

The obvious starting point for estimating the deliverability of a well, a reservoir, a field, or a group of reservoirs or fields is an estimation of reserves. Since for many purposes deliverability estimates must be as certain as possible, the concept of "proved reserves" is a useful one. This would certainly be the case if estimates were being made for such purposes as the financing indicated in the example above. However, since the deliverability concept has a time dimension, it may be of critical importance to have the best possible estimate of the *duration* of specified delivery rates *and* the *probabilities* of these rates declining in the future. This type of information must rely heavily on *probable* reserves estimates as well as on proved reserves estimates.

A further consideration arises from the fact that a contract for interstate deliveries by a gas producer from a field, well, or reservoir may constitute "dedication" of all the gas in the field or reservoir. This may be true even though the contract between producer and pipeline has a termination date. The producer, having received a "certificate of public convenience and necessity" from the Federal Power Commission to sell gas in interstate commerce, must also get FPC permission to terminate deliveries or to change the price. A pipeline will have many contracts with owners and part owners of wells and fields, each contract taking cognizance of different conditions—physical or economic—that exist for each producer. Multiple ownership of reserves and production complicates the task of both the pipeline and the regulatory body. Deliverability and the total *ultimate* recoverable reserve thus become highly significant for the pipeline, the producer or producers selling the gas, and the regulatory agency. The "proved reserves" concept may be quite inappropriate, particularly where the drilling of development wells is, in part, contingent upon the gas purchase contract being negotiated. What seems to have occurred in the case of gas reserves estimates for interstate sales is to depart, where necessary, from the "proved" concept. As was noted earlier, in the case of gas, total recoverable reserves are often easier to estimate than in the case of crude oil; therefore, the "proved reserves" figure may often be close to the total recoverable reserve figure.[25]

A particular difficulty in estimating gas deliverability is encountered in working with "casinghead" gas. For example, in the Permian Basin Area Rate Hearing, a total of 89 reservoirs (out of a total of 2,350 reservoirs in the Basin), which accounted for about 75 per cent of the total reserves under consideration, was the sample taken to estimate deliverability for the entire area. Fifty-four of the 89 were casinghead gas reservoirs. Casinghead gas accounts for about two-thirds of total gas reserves in the Permian Basin. In estimating deliverability, it is necessary to predict oil production, which is partly a function of (1) oil proration regulations and (2) maximum gas-oil ratios allowed by the regulatory authority. The problem is further complicated by pressure maintenance projects utilizing gas or water injection and by production schedules of producing companies which are concerned primarily with oil rather than gas. Oil proration can be a particularly large variable in gas deliverability. In 1948, most oil wells in Texas subject to proration produced 366 days during the year; in 1956 they produced 190 days; and in 1960, 104 days. Such wide variations have a tremendous impact on annual gas deliverability and ultimately on recoverable gas reserves, although reserves are not affected to as great an extent as deliverability. However, to the extent that field production and development depend on payouts, which in turn are based on revenues from oil and gas, continued suppression of production may tend to slow down or halt development and hasten abandonment.[26]

The previous discussion of gas reserves and availability data casts some doubt on the usefulness of the data with respect to company decision making and public policy making. The deliverability concept, which is in part a function of reserves, brings into the discussion many of the technical and institutional factors which must be considered in making a gas reserve usable. While it is beyond the scope of this study to explore the existing work being done in this area, it is important to note the need for further statistical and analytical efforts. It seems likely that more and more emphasis will be placed on good estimates of deliverability by company, region, and the nation in the regulatory work of the FPC, if regulation continues along the lines currently contemplated by the FPC.

[25] See *ibid.*, Testimony of C. E. Turner, Transcript pp. 242-71, for a discussion of gas reserves calculations for FPC purposes.
[26] Gardescu, *op. cit.*, contains an interesting discussion of the problems of estimating deliverability. See also a series of four articles by Henry D. Ralph, on "Crisis in Gas Proration," *Oil and Gas Journal*, January 14, January 21, January 28, and February 4, 1963, for a good discussion of the extremely complex problems relating to gas production, delivery, reserves and regulations.

PROBLEMS OF COMBINING CRUDE OIL, NATURAL GAS, AND NGL RESERVES DATA

While it is true that crude oil, natural gas, and NGL are of the same general hydrocarbon family, the problem of reducing these different raw materials to some common denominator is a difficult one. At the same time, such a statement in terms of a common denominator would be useful when long-run supply, demand, availability, capacity or cost studies are being considered.

One method of combining reserves is to equate them on a Btu (British thermal unit) or heat content basis. Using average estimates of heat content, 1 million barrels of liquids is equal to about 6 billion cubic feet of gas. Mr. Lahee, in estimating the reserves found from exploratory drilling effort, uses this technique of aggregation.[27] This method of stating reserves is far from satisfactory because one million Btu's of energy in the form of crude oil or NGL does not have the same value at the well head as 1 million Btu's of energy in the form of gas. In general, liquids are several times more valuable than gas at the production site because of the uses to which they are put and the lower cost per Btu of getting liquids to market.

An alternative method of combining reserves is to use a value basis, for example, $2.90 per barrel of liquids and 15 cents per mcf of gas. However, when a value basis is used the ratio of the quantities of reserves of gas relative to liquids reserves changes as relative values change. If, for example, the price of gas rose relative to oil, gas reserves thus measured would rise relative to oil reserves. It should be made clear that under the value method, there may be a shift in the relative proportions of oil and gas reserves in the combined total when relative values change, and this may occur without any change occurring in the actual quantities of either. Looked at from the viewpoint of a company making investment choices between exploration for and development of gas or of oil, the relative value of reserves becomes highly significant.

While this problem need not detain us here, it is perhaps important to note that the problem of aggregation is one that needs solving because of the frequent desire to speak of total hydrocarbon reserves. Also, in cost studies such as those done in FPC proceedings it is essential that some method of aggregation be used. A standard method is needed by which to add crude oil, NGL, and natural gas reserves. The result will be to some extent arbitrary, but need not be misleading if properly explained and interpreted.

[27] See Chapter 5 for a discussion of Lahee's data.

III

SUMMARY AND CONCLUSION

12

Summary View of "Reserves Estimates"

In bringing this study to a close, we shall first summarize the information now available on reserves; then state the considerations of public interest which dictate an expansion of this information; and finally, suggest directions in which the expansion of information might proceed. The present chapter deals with the first of these topics. For convenience, we shall conduct the discussion in terms of crude oil, but what is said is, for the most part, *mutatis mutandis*, applicable to natural gas and natural gas liquids.

Estimates of "reserves" may be likened to a set of concentric circles, widening as restrictive assumptions are dropped and new considerations admitted. In summarizing it will be useful to go back over some earlier material to show by what stages the circle expands.

1. At the center, as the smallest circle, are the American Petroleum Institute (API) estimates of proved reserves, which at the end of 1964 were 31.0 billion barrels (or 38.7 billions if natural gas liquids are included). Every informed person knows that, even under relatively fixed economic and technological conditions, the amount of recoverable oil is much larger than this. The API estimates provide an annual statement of the "working inventory" of crude oil in a closely defined area underlying or contiguous to wells already drilled. The estimates are based on restrictive assumptions concerning geographical coverage, geological knowledge, recovery technology, and economic conditions. Since the purpose of the estimates is specifically to measure this firm inventory, it is no criticism of the figures, as such, that they do not perform the function of estimating the future availability of oil in some broader sense. To perform the latter function, supplementary methods of estimation are required.

147

While limited in their scope and purpose, the API figures do provide interesting and useful information. In particular, the course of proved accruals to the inventory can be checked against the rate of withdrawal through production. Experience also provides certain rule-of-thumb relationships, such as that extensions and revisions average roughly 5 to 6 times the amount annually added to proved reserves from new discoveries.

2. The first expansion of the circle comes from the very nature of the API estimates. Future extensions and revisions are bound, in the aggregate, to increase the later proved contents of known fields. The most recently discovered are the most severely understated in current estimates, but additions due to extensions and revisions continue for long periods of time from the original dates of discovery of fields. Upon the basis of expert opinion, a tentative value of 25 to 35 billion barrels of crude oil can be placed on this expectation. The formula used by the Department of the Interior for the McKinney report is that future extensions and revisions of the API estimates will be equal to the amount of proved reserves as stated at any given date. The task force survey of industry opinion for the same report yielded a somewhat lower multiplier.

This is the point at which it can perhaps be said that the informational sources provided by the industry begin to show their inadequacy. Individual industry experts are quite willing to talk about reasonable expectations in this sphere, and company officers occasionally say something about them in public speeches; but there is no established avenue through which collective expert judgments, or ranges of judgment, are given public expression. Because of the uncertainties involved and difficulties of anything like precise determination, it is understandable why the API should hesitate to enter this field of estimation "officially." At the same time, there is no necessary barrier to some regular, if unofficial, discussion of these anticipations from industry sources.

3. A different approach to widening the circle around the center of proved reserves is that followed by the Interstate Oil Compact Commission (IOCC) reports in estimating the additional reserves which could be economically recovered by the application of present conventional secondary recovery methods to existing reservoirs. The 1962 IOCC report gives this a value of some 16 billion barrels. This type of estimation is designed to expand the field of firmly established reserves beyond the restrictive definitions of the API estimates, a perfectly proper purpose. At the same time, an element of confusion is introduced. Any later development of fluid injection opportunities will be included in later revisions of API proved reserves, and therefore constitute a portion of the expanded expectations referred to in item 2 above.

In any effort to expand the range of estimation, the possibilities presented by secondary recovery and pressure maintenance should certainly be identified and receive special consideration. A reasonable procedure would provide that such studies be integrated within a comprehensive plan, rather than leading a separate life of their own as in the past.

4. The next expansion of the circle comes from conjectural results of improvements in the technology of recovery which would increase the percentage of anticipated recovery from the oil originally in place in known fields. No one can, within any close range of accuracy, actually "estimate" the results of methods as yet unknown, untested, unapplied, or inapplicable under present economic conditions or regulatory practices. At the same time, substantial results are anticipated from these sources, and some account must be taken of them in any consideration of the future availability of oil. The measurement, either of potentialities from these sources or of results achieved, depends upon the existence of a figure for original oil in place, against which actual or anticipated recovery can be calculated as percentages.

The IOCC has provided such a figure (the latest [1962] being 346.2 billion barrels in known fields); and, as we have seen, this figure is seized upon by almost everyone who wishes to project the potentialities of oil supply. The reason is easy to see: a whole range of facts and anticipations can be quantitatively compared. For example, crude oil produced to 1962 (67.8 billion barrels) can be stated as 19.6 per cent of oil originally in place. Add in API proved reserves (31.8 billion barrels), and the percentage becomes 28.8. Add in anticipated future additions to proved reserves in known fields of 30 billion barrels, and the percentage rises to 37.4. If 50 per cent could be achieved, the result would be 173 billion barrels; and, after deducting oil already produced, would leave 105 billion barrels still to be recovered from known fields. Given our 31.8 billion barrels of proved reserves, every percentage point improvement in recovery would add nearly 3.5 billion barrels of reserves.

This is a nice simple way for analysts to prepare figures which can be easily digested by planners and policy makers; and it is perfectly valid if the oil-in-place figure is credible. As we noted earlier, there is some skepticism in professional circles about the credibility of the IOCC figure. Nevertheless, it might be regarded as useful by those who use it, even if it had a margin of error of 50 billion barrels in either direction, making the figure range anywhere from 300 to 400 billion barrels. In a field beset with great uncertainties at best, it at least provides a home base from which to think quantitatively about problems of future oil supply.

We are not ourselves convinced that the kinds of calculations based upon a single oil-in-place figure represent a satisfactory approach to quantification of the various potentialities of future oil supply. The figure lends itself to the criticism, expressed by some industry experts, that it adds pears, peaches and apples into a category of fruit. However, some basis for quantification is needed; and, if a better one is to be had, presumably the source will have to be the industry itself. Possibly the forthcoming report of the API Subcommittee on Recovery Efficiency could provide a starting point for a reconsideration of this problem.

5. Up to this point we have been speaking of oil reserves in known fields. The final stage in the consideration of future availability of oil is the attempted quantification of reserves from oil in fields yet to be discovered. The range of speculation is wide in the face of two unknowns: (1) the quantity of oil in place to be discovered and (2) the percentage rate of recovery. Bruce C. Netschert, as we saw, on some sort of averaging of other estimates, arrived at 235 billion barrels of oil in place to be discovered. C. L. Moore, by a projection of statistical trend lines, arrived at 158 billion barrels. The range of other figures is wide. The "reserves" to be computed depend upon the percentage rate of recovery applied. The forecasting in these areas naturally divides itself into two parts: (1) the short-term prospects and (2) the long-run, or "ultimate," prospects. With respect to the first, given an improved base of statistical series, it should be possible to extrapolate a trend, or range of trends, of oil discovery in relation to discovery effort, and thereby to provide a forecast with some claim to credibility—always subject to revision on the basis of accruing experience. With respect to the second, forecasting appears to have little purpose except to generate lines of thought concerning public policies relative to various possible contingencies in the field of energy requirements and supply.

Items 1 and 2 in the preceding list present highly credible orders of magnitude, in the region of 60 billion barrels, in relation to known fields and without striking changes in technological and economic conditions. Item 3 largely overlaps item 2. Items 4 and 5 open up a wide realm of speculation, though they undoubtedly point to substantial additions to future reserves through improved recovery and new discoveries. In addition to the physical and technological factors, economic factors will play a large part in the outcome. If "the price is right," an intense search for new oil will yield results, and methods of increasing yields will be more extensively introduced. Import controls, state regulatory practices, competitive sources of energy (tar sands, oil shale, coal), and the technology of energy utilization are all parameters of the problem.

13

Aspects of the Public Interest

The public interest in future domestic availability of petroleum needs no explanation. Petroleum is at present the main source of energy for the industrial system and a vital element in the system of national security. The amounts which may ultimately be recoverable from the earth cannot be estimated within any reasonable orders of magnitude. The supply, while very large, is also finite; and the amount which is visible in the form of measurable or inferable reserves would carry through only a short period of time. It is therefore unnecessary to argue the fact of public interest in information concerning the future availability of petroleum—the "public" interest referring not just to agencies of government, but to everyone dependent on the industrial system, notably including oil companies.

INFORMATIONAL NEEDS

So much having been said, it is not immediately obvious what sorts of information are needed for what purposes, and to what extent it is feasible to provide them. There is no limit to the detailed information which public agencies might at times desire in fulfilling their responsibility for formulating policies directed toward the future adequacy of supplies of energy. Practical questions arise: Is it obtainable? If so, how important is it? At what cost can it be made available? And is it worth the cost? These questions, it will be seen, are not so different from those raised inside individual companies as to their expenditures on reserves information. But the context is different. For companies, the purpose is to serve management

151

decisions directed toward profitable operation. For government agencies, both state and federal, the purpose is to serve policy decisions directed in some broad way toward long-run economic well-being and national security.

The widely circulated (January 1965) report by the Interior Department, *An Appraisal of the Petroleum Industry of the United States*, presents an excellent example of the degree to which a major oil policy agency is hamstrung by research data inadequacies. This report, heralded as a definitive study on the posture of the domestic petroleum industry, concludes: "There is, then, a need for the development of additional reserves and an equal need to conduct such development in a manner which does not contribute to already excessive producing capacity."[1] Yet the discussion of reserves is limited to reviewing API-AGA and IOCC proved reserves figures, generalizing that "Published estimates . . . of the (ultimate) recoverable crude oil resources in the United States range from about 200 billion barrels to about 600 billion barrels,"[2] and to giving warnings that economic incentives are needed to convert resources into reserves.[3] It is unable to analyze in depth the impact of past technological and economic factors on reserves or to make significant comments as to the specific forces that will affect reserves in the future and to what degree. Clearly, the question asked by the report—how to increase long-run reserves without increasing productive capacity—requires much better reserves information than we now possess.

Private companies are much better served with information for their purposes than government agencies are for theirs. Companies not only estimate their own proved reserves, but are also in an informed position to speculate reasonably upon the "probabilities" and "possibilities" of their properties, to check results against earlier estimations, and to revise their judgments on the basis of experience. With some companies, at least, this type of analysis is carried out in great detail in ways which can be aggregated and traced through time as correlations of investment outlays with discovery results.

Nothing comparable is available to government agencies. From the API they get the annual statement of gross additions to reserves and changes in cumulative net reserves. That is all they get "officially" from the indus-

[1] U.S. Dept. of the Interior, *An Appraisal of the Petroleum Industry of the United States* (Washington: 1965), p. 44.

[2] *Ibid.*, p. 13.

[3] The report does reveal new estimates by the Geological Survey of oil originally in place. See *supra*, p. 65.

try. The IOCC has added something in the way of estimated secondary recovery possibilities. Beyond this, industry experts can be informally polled to get their opinions as to how much recoverable oil they believe is likely to be proved up in presently known fields. That is about the whole story. Beyond this, anyone can start speculating upon possible percentages of recovery and upon discovery possibilities, which is not without some usefulness, but not very informative.

Since there is a mounting dissatisfaction in government circles with the limited informational basis for the discussion of future petroleum availability, the industry is under pressure to expand its reporting services. There is, indeed, a "soul-searching" API group considering what may be done to satisfy claims upon the industry; and there is also an inter-agency government group trying to determine what should be claimed. These informational discussions are much wider than the subject of reserves, but we limit ourselves to that aspect. Since some changes in the situation will presumably take place, we may usefully look at some issues which have to be resolved.

The first question to be raised is why government agencies need—or think they need—more information. Among a group of "oil men," anyone can start an argument just by asking that question. They do not all think alike, so it is a real argument among themselves, not just between themselves and "outsiders." On one side of the argument is a conventional attitude, widely held in the industry, that the federal government need not worry about problems of future energy supply. All government needs to do is to maintain adequate incentives to exploratory activity and technological advance in recovery. There is plenty of oil and gas in sight for the near future and, if it begins to run low, technical ingenuity under private enterprise will be turned to the exploitation of oil shale, tar sands and other alternative sources of energy.

This philosophical position, expressing extreme distrust of intrusions by the federal government into industry affairs, does not square easily at all with certain relations between government and the industry which limit the force of the "private enterprise" argument. The domestic producing industry operates under the umbrella of federal import quotas—a thoroughly unorthodox procedure under American commercial policy—based upon the presumption that such protection of the industry is necessary for reasons of national security. Further, the industry is accorded unusual tax treatment on the presumption that the search for petroleum is important and will only be carried on adequately if supported by this stimulus. Moreover, the domestic production rate and price are stabilized by the action of state regulatory commissions. An industry so thoroughly dependent upon public action for its well-being is not in a good position to deny the right of government to be well-informed concerning the industry's operations and prospects. Beyond this, it is hardly conceivable that any responsible government could fail to interest itself in the fundamental problem of future availability of energy.

Assuming, then, that the federal government has adequate reasons for desiring improved information on petroleum reserves,[4] the critical question is the kind of information which government agencies need and the extent to which these needs can be met by reasonable demands upon the industry. The basic need, it appears to us, is to be able to project the *unfolding* pattern of supply. Evidence is needed which will throw light continuously on what is *emerging*. The most important evidence of this sort would appear to be statistical evidence which would show (1) trends in discovery relative to discovery effort and (2) trends in the rate of recovery from discovered oil and gas in place.

By this test, the speculations about "ultimate" reserves become of secondary importance—though not useless because they identify problems that will have to be met sooner or later, and establish parameters within which to think about the role of petroleum in relation to other sources of energy. No one can predict at what rate the finite supplies of recoverable domestic petroleum will fade away. Some say, "faster than you think." Others are more optimistic. The informational problem, from a public point of view, is primarily that of the ability to discern trends relating to rates of discovery and recovery.

DISCOVERY TRENDS

A great informational advance would be achieved by a logical extension of the present API reporting procedures. Letting the present reports continue unchanged, the same basic data could be used to build up a supplementary record by attributing each year's "extensions" and "revisions" back to the year of discovery. Such records, kept year after year, would reveal trends in the discovery of oil which, when correlated with the intensity of exploratory effort over time, could be highly informative. To avoid any ambiguity in the meaning of "trend of discovery," the

---

[4] State government agencies, particularly taxing and conservation bodies, also have needs for such information, although these needs, thus far, have usually been voiced privately or informally rather than publicly.

concept we are using is that the exploratory well which reveals a new field "discovers" all the oil credited to that field by later development. Currently, the API furnishes a static or "stock" concept of reserves. The dating-back process would, as time passed, provide a dynamic or "flow" concept which for many industry and public policy questions would constitute a far more useful tool of analysis. The charting of the data would reveal unpredictable peaks and hollows. But a calculated trend line, correlated with discovery effort, would support certain reasonable expectations concerning the oncoming supply of oil.

It is, of course, true that data recorded in this fashion become increasingly deficient for the most recent years. Fields with a productive life of 20 years have the back-dated assignment of 20 years of "extensions and revisions," those of 5 years life have only 5 years of such assignments. Because of this lag element in "proving" reserves, it is known with assurance that the reserves underlying the more recently discovered fields are very much larger than the amount reported as proved; the question is, how much larger.

This deficiency of present knowledge is capable of being overcome to a substantial degree by refined statistical analysis. If, in every year, the extensions and revisions of that year were attributed back to the year of discovery, it would be possible, after a certain lapse of time, to derive a factor which could be used to blow up the initially proved "discoveries" for a particular year into figures of probable eventual recovery (within limiting economic and technological assumptions). The factor could not be applied to individual fields, and might indeed be substantially off for the total discoveries of any particular year; but by the logic of large numbers it should give credible results. The factor itself would be subject to revision with accruing evidence. This method of blowing up current discoveries has inherent limitations in the economic and technological assumptions which are applied. These limitations could, however, be overcome by progressive change in the factor, and by special consideration of reserves based on fluid injection or technological advances in recovery. *Ranges* of future reserves could at any time be secured by applying alternative economic and technological assumptions.

The procedure based on back-dating has the practical merit that it ties in with the established API system of reporting. Another type of approach is sometimes suggested, based upon procedures used in the reserves analysis of some individual companies. In Chapter 5, it will be recalled, we presented reserves classification plans by Lahee, Arps, and DeGolyer and MacNaughton, a common feature of which was that they used the categories of "proved," "probable," and "possible," applied to reserves divided between "primary" and "secondary." By applying such a classification, companies could if they wished—and some do to a degree— surround each discovery well by concentric boundaries of "proved," "probable," and "possible" areas, assigning speculative amounts of "reserves" to each. As development proceeded, periodical revision could assure that the boundaries be relocated, the flow of "reserves" from one category to another be recorded, and the total be revised.

A system like this could produce aggregate regional and national figures if applied to unitized reservoirs under standard definitions and procedures, and the aggregate figures could be turned into trend lines over time. Something of this sort is done in the Soviet Union, where the process is made feasible by government ownership. "Reserves" are classified into A_1, A_2, B, C_1, and C_2, as noted earlier with reference to natural gas (see Chapter 11). $A_1 + A_2$ corresponds roughly to API proved reserves, drilled and undrilled. The other categories are described as follows in a Census study:

> For category B, there must be favorable sediments, well logs, and commercially significant flows of oil from at least two wells in the field. Analysis of the oil and gas must have been made, and the general structure of the deposits established. However, the distribution of productive strata, the properties of the collectors, and allied data need be known only approximately. Category C_1 generally includes new fields, and new pools in existing fields, as well as undeveloped portions of structurally heterogeneous deposits. Category C_2 reserves are based on the expected resources of the strata which have proved productive in the same oil and gas province, and on favorable geological and geographical indications.[5]

The thought naturally arises: Why cannot some similar system be applied to American reserves, using a classification like, for example, that of J. J. Arps (see Chapter 5). Suggestions of this sort bring to focus the question of what is feasible and what is not. Our discussion in Chapter 5 of the nonuniform character of reserves analysis by petroleum companies indicates the reason why this plan is not immediately feasible. Each company uses its own method, and there are no homogeneous statistical units to be aggregated. Moreover, the fact that pools are not unitized but are administered under fragmented leases adds to the difficulty of this type of estimation. If the matter were considered of sufficient importance, the

[5] Demitri B. Shimkin, *The Soviet Mineral-Fuels Industries, 1928-1958: A Statistical Survey*, Bureau of the Census, U.S. Department of Commerce, 1962, p. 159.

government could, of course, attempt to impose a uniform method. (It did impose uniform accounting upon the railroads.) To do so, however, it would be necessary to reconstruct the whole structure of the industry—a high price for a modest informational advance. Back of the statistical obstacles lies the institutional fact that, in the highly competitive petroleum producing industry, the companies are unwilling to share fully their knowledge or opinion of the undeveloped potentialities of properties in which they are interested. Attractive as the idea is, we do not see any basis, in information possessed by producing companies, for aggregated regional and national estimates of potential reserves in the categories of "proved," "probable," and "possible."

While the obstacles just stated are no doubt insuperable for anything "official" like API estimates, some avenues are open for improvement. One possibility is for the industry, alone, or in co-operation with interested state and federal government agencies, to work on standardization of definitions of the various reserves categories for internal company use. It is unlikely that unanimity among companies could be reached on the several concepts, but at least there would likely be a narrowing of the current differences and mutual enlightenment among the companies on these problems. If some standardization of concepts such as proved, possible, and probable—or whatever might be substituted for any of these—could be accomplished, individual companies possibly could report to a competent government agency whose function would be to collect, aggregate and disseminate information in a form which would not reveal individual company data.

A second possible avenue for improvement, and one less ambitious than the first, would not attempt to standardize definitions or concepts, since this alone is a substantial barrier. Instead, companies which make any kind of reserves estimates might make the results and the specific definitions used available to a competent government agency. Such material, though incapable of being aggregated statistically, might be very informative as evidence for expert assessments of parts of the problem.

To conclude this section, we revert to the beginning of it where we suggested that useful trend analysis on discovery could be derived from the continuous annual back-dating of "extensions and revisions" of proved reserves to the year of original discovery of the fields in which they are located. A start has already been made in the National Petroleum Council studies. Though we are aware of certain technical and practical difficulties, this appears to us the most natural route for expanded API reporting to take, and the one most likely to yield useful trend information. We would add that fuller discussion of definitions, procedures, and limitations would make such series more useful for analytical purposes.

TRENDS IN RATES OF RECOVERY

An important factor in the future availability of petroleum will be the degree of efficiency in the recovery of oil. A substantial addition to reserves from this source is anticipated; but there are at present no statistical data from which to estimate the degree of improvement. To a limited extent, the "revisions" introduced into the annual API estimates reflect this factor, but in a manner which makes it impossible to isolate its quantitative effects. If the annual revisions were broken down into (a) "revisions from secondary recovery and pressure maintenance," and (b) "other revisions" —each being credited back to the year of discovery of the field—a considerable body of additional knowledge would begin to emerge. Difficulties of classification would no doubt be encountered, especially where pressure maintenance is introduced early in the life of a field or pool, but even so the less-than-perfect addition to knowledge should be considerable. Where unit operation is introduced early in the life of a reservoir, possibly the effect upon the estimates could be isolated.

A special difficulty in evaluating improvements in recovery is that they can be measured only by reference to some estimate of oil originally in place. In the literature of the industry there are many off-the-cuff statements about the average recovery percentage for the country as a whole, but we are unaware of any publicly available evidence upon which they are based. Individual companies are in a position to assess the performance of their own properties, and some of the larger ones are probably in a position to generalize somewhat from their own experience concerning recovery potentialities. But the records of companies provide no basis either for aggregated reporting on the past degrees of improvement or for measuring its trend.

As we have seen, the IOCC estimate of oil originally in place has been commonly used as a basis for certain types of calculation, purporting to express percentage rates of recovery. It can, for example, be divided into cumulative production, plus proved reserves, plus currently unproved or unrecoverable oil; or into other numerators otherwise arrived at. Even if the IOCC estimate of oil in place were considered a sound figure—about which doubts are widely expressed in the industry—there is still no established procedure for periodical estimates of the amount of oil which may reasonably be expected to be recovered. In the absence of any procedural apparatus for such estimation, it is difficult to see what credence can be given to statements about either current rates of recovery or the trend in such rates. The kinds of calculations based upon the IOCC estimates have afforded a sort of solace to those who craved for some sort of light in this dark corner; but even that solace will be removed by the announced abandonment of the IOCC series.

The methods by which to create new light in this area are not at all self-evident. Not only is oil in place a very troublesome figure to pin down, but rates of anticipated recovery are different for all sorts of different reservoirs; detailed data on reservoir performance are not generally available for aggregative treatment; to the extent available, they are difficult to reduce to an index number of performance; and estimates of recovery are heavily dependent upon economic assumptions. These obstacles create a good deal of skepticism within the industry about the feasibility of providing the sort of information which government agencies might ideally like to possess.

We, ourselves, are not convinced of the force of this skepticism. We recall that forty years ago the concepts of Gross National Product and National Income were little beyond the stage of statistical dreams; now they are the foundation of all discussions of national economic policy. Statistical science is a very powerful instrumentality which can make unprecedented advances if the need for them is felt with sufficient urgency. We referred in Chapter 4 to the present investigations of the API Subcommittee on Recovery Efficiency. It is conceivable that its work might be the opening stage for a great advance. We suspect that the informational situation of the future will not be one in which precise percentages can be derived from a neat figure on oil in place and a neat estimate of anticipated recovery; but that policy making will require a mastery in greater detail of the complexities of the informational situation. There still lies upon the industry and government agencies, jointly, the problem of devising the system for providing the information, within limits of practicality. "Practicality" is a very elastic word, dependent for its meaning upon the intensity of the desire to receive, on the part of government, and the degree of willingness to give, on the part of the industry.

TENTATIVE GOVERNMENT PROPOSALS

The primary sources of improved information on reserves will necessarily be industry sources. Suggestions for expanded reporting have encountered a number of obstacles which have been noted at earlier points in this study:

1. Some of the information possessed by companies is confidential to their own competitive purposes.

2. Because of nonuniform methods of recording company data, there is no convenient way of aggregating such data as there are.

3. Companies are generally not interested in incurring costs beyond those which have a direct bearing on management decisions.

These obstacles, while real enough, for they place limits on some kinds of information which might ideally be desired, are by no means serious enough to prevent substantial expansion of reserves data. The API, for example, has opportunities for extending its system of reporting in a number of informative ways. To the list of inhibiting factors should possibly be added a general air of suspicion toward federal agencies when they begin to exhibit an intrusive interest in the affairs of the industry. There is a long history of resistance to federal regulatory encroachment, and even such apparently innocent things as expanded data on reserves and productive capacity may appear like the camel's nose under the edge of the tent.

Given the inhibiting factors within the industry, the initiative for an expanded system of reports will presumably have to come from government. Such a system cannot, however, simply be invented and imposed. The possibilities will have to be explored with industry experts in order to arrive at some meeting of minds about what is feasible and what the industry can reasonably be expected to do.

Such initiative has in fact been undertaken by interagency groups of government technicians, and has led (March 1965) to a Petroleum Statistics Report, prepared by the Petroleum Statistics Study Group, chaired by the Bureau of the Budget in the Executive Office of the President.[6] A draft of this report was made available to all interested parties for critical comment prior to the preparation of a final report. In addition to the section on "Reserves," the draft contains sections on "Productive Capacity," "Wells Drilled," "Transportation and Deliverability," and "Expenditures and Revenues."

The categories of the proposed reporting framework on reserves is shown in the following table from page 10 of the report.

---

[6] The study group consisted of representatives from the Departments of Commerce, Defense, Interior, Justice, State, Bureau of the Budget, Council of Economic Advisers, Office of Emergency Planning, and the Federal Power Commission.

Cumulative Production and Classes and Categories of Petroleum Resources in Known Fields.

Classes:	A. Cumulative Production and Defined Resources*	B. Indicated Resources
Categories:	1. Cumulative Production	1. Recoverable
	2. Proved Recoverable Primary Reserves	
	3. Proved Recoverable Secondary Reserves Currently Producible	
	4. Additional Secondary Reserves Recoverable by Methods Currently Economic	
	5. Additional Secondary Reserves Physically Recoverable by Known Methods	
	6. Unrecoverable	2. Unrecoverable

* Equals petroleum originally in place in the defined portion of the field.

To master the full meaning of the table and associated reporting system, it will be necessary for readers to read the entire section of the report dealing with reserves. The general character of the categories may, however, be simply explained.

Column A, "Cumulative Production and Defined Resources," is constituted by combining the reporting categories heretofore used by the API and the IOCC, as follows:

(a) The total of items 1–6 corresponds to the IOCC "oil originally in place in known fields."

(b) Items 2 + 3 correspond to API "proved" reserves, and equally to IOCC "primary" reserves, plus a portion of IOCC "secondary" reserves.

(c) Items 3 + 4 correspond to IOCC "secondary" reserves based on conventional fluid injection methods under existing economic conditions.

(d) Item 5 is the IOCC category described in the same words.

(e) Item 6 is estimated original oil in place minus items 1 to 5.

Column B, "Indicated Resources," is in effect a currently estimated amount of oil in place in presently known fields, but outside the currently defined limits, with a further estimate of recoverable oil therein. This amounts to estimating the "probable" total and recoverable content of the unproved portions of reservoirs. As they are defined geographically, the contents would move into the appropriate categories of Column A.

It is proposed that such reports be made on each field and/or reservoir, in a manner to facilitate aggregation of the figures into regional and national totals. Changes in the figures would be due to "discoveries," to "modifications" which changed the estimate of oil originally in place, and to "shifts" among the categories. Data changes for each year would be recorded back to the year of discovery of each field as the basis for analysis of discovery trends.

It will be seen that this draft report by the Petroleum Statistics Study Group brings into sharp focus the problems of estimation outlined in our study, and is designed to integrate a total system, in place of the partial and spotty forms of evidence now available. It raises difficult questions of feasibility which will have to be thoroughly discussed. In whatever ways it may need to be modified, it presents a challenge to the industry to work with public agencies in improving the informational basis for estimating the future availability of petroleum. Our discussion of the report has run in terms of crude oil, but it is designed also to cover the subjects of natural gas and natural gas liquids.

THE ROLE OF A GOVERNMENT AGENCY

During the same period that (in some quarters of government) the industry has been criticized for the paucity of information it provides on reserves, the federal government has been negligent in arranging for the collection and analysis of such information as there is. It seems almost inconceivable that, when the Senate Interior and Insular Affairs Committee and the Joint Committee on Atomic Energy wished to assemble material on the future availability of hydrocarbons, they found it necessary to rely on hastily assembled groups to make *ad hoc* surveys. One would suppose that, in a country concerned for its future supplies of energy, there would be a specialized agency or group, not merely to accumulate data, but to apply sophisticated analytical procedures to such data as could be assembled.

Even now, there are sources of information in the industry which could be mobilized much more effectively than they have been in the past. Most of the existing knowledge concerning reserves and concerning the potentialities for recovery is contained in the minds and in the files of a relatively small number of expert personnel within a few large companies and consulting firms. Part of it is based upon systematic estimation within their companies, part of it upon knowledge acquired more informally from the geological and engineering personnel of their companies. Experts need not agree, and the information could not be pooled into neat forms of statistical generalization, but it could be systematically reviewed. If an agency were designed for this purpose, it would need a better basis of "diplomatic relations" with industry sources than has heretofore existed.

Other potential sources of information for such an agency lie in the expert personnel of state and federal agencies. Also, the voluminous records, often on a field-by-field basis, in the hands of state conservation authorities and the Federal Power Commission, represent a vast unmined source of information on reserves and producing capacity. While some state regulatory agencies have taken advantage of their unique opportunities for assembling data on reserves, others have been negligent. The possibilities from this source would be much enhanced by having common statistical standards which agreed with those to be used by industry sources and federal agencies.

Apart from making the best of what is now available, any substantial improvement in knowledge will depend upon expanded reporting and improved analytical procedures. The more elaborate the data system, the more important will become the quality of the analytical skills to be applied. There are perhaps several ways to get this task done. If a government office were established, an expanded informational system would require a specialized staff whose duties included the collection of data from all sources, systematic analysis and synthesis, and periodical reporting of the results in easily understood form. The exact scope and role of such an office could, however, be greatly affected by attitudes within the industry. If industry agencies were willing to undertake a large-scale improvement of the statistical data situation, they could minimize the measures which government agencies might otherwise feel it necessary to undertake.

Methods of Estimating Reserves of Crude Oil,
Natural Gas, and Natural Gas Liquids

BY WALLACE F. LOVEJOY AND PAUL T. HOMAN

designer:	Athena Blackorby
typesetter:	Baltimore Type
typefaces:	Bodoni #375, Times Roman
printer:	John D. Lucas
paper:	Clear Spring Antique
binder:	Moore & Co.

Printed and bound by CPI Group (UK) Ltd, Croydon, CR0 4YY

22/10/2024

01777605-0004